Fire and Rain

Stephen LaDochy · Michael Witiw

Fire and Rain

California's Changing Weather and Climate

 Springer

Stephen LaDochy
California State University
Los Angeles, CA, USA

Michael Witiw
Sammamish, WA, USA

ISBN 978-3-031-32272-3   ISBN 978-3-031-32273-0   (eBook)
https://doi.org/10.1007/978-3-031-32273-0

Cover credit: Kornei/Stock.adobe.com (rain), toa555/stock.adobe.com (fire)

This Springer imprint is published by the registered company Springer Nature Switzerland AG
The registered company address is: Gewerbestrasse 11, 6330 Cham, Switzerland

For my wife, Beth, our children Mike and Lisa, and our grandsons, Henry, Harry, Lewis, Eddie, Clive and Ben for their love, support, and encouragement.

Michael Witiw

To my wife and daughters, Sue, Lisa and Jenette, who wondered why I'm always looking skyward.

Stephen LaDochy

Foreword

All you weather and climate wonks, *Fire and Rain: California's Changing Weather and Climate* is not to be missed. This well-researched book is the latest important contribution to understanding California's diverse and dynamic meteorology and important climate trends as well as related state water issues and natural disasters. Stephen LaDochy, long-time faculty member of the Department of Geosciences and Environment at California State University, Los Angeles and instructor of meteorology, and Dr. Michael Witiw, Certified Consulting Meteorologist and instructor of meteorology, have compiled a timely, very informative, and easy to understand tutorial on the wide-ranging weather and climate phenomenon of the Golden State.

In the Los Angeles area, for example, the weather is an everyday topic of conversation, speculation, and even humor. Remember Steve Martin's take on Los Angeles weather as the wacky TV weatherman, Roland Harris, in *LA Story*, "if you turned off the sprinklers, LA would turn into a desert." Funny, but it's a one-dimensional, simplified view of our weather. Yes, it often seems that the Los Angeles area weather hardly changes from one week to the next. But in fact, the weather varies from benign to some of the most violent events found anywhere on the planet. From season to season, the LA region can whiplash from months of meager rainfall punctuated by brutal heatwaves, to fires that scorch much of our landscape, to extreme rainfall accompanied by floods and mudslides. All these events are modified by the extreme range of California's terrain, from the coast where the great Pacific Ocean acts to limit the range in temperatures to great mountain ranges that dominate the California landscape and give birth to extreme Santa Ana wind events or overwhelming rain storms. Our climate ranges from months of dryness and heat to violent wind, rain, and flooding episodes. The daily life of Californians is governed by this rich meteorological mix. California's vast agricultural industry, our complex transportation systems, especially the freeways, our gigantic ports, and how we plan our daily lives, are controlled and often disrupted by the weather. That is why the most watched segment of the 6 pm Evening News is the weather. All Californians think they are weather experts.

Fire and Rain: California's Changing Weather and Climate provides a clear explanation of the broad range of California weather phenomenon. The often-complex scientific explanations of these weather events are presented in a manner that will be easily understood by interested aficionados, students and appreciated by professionals. The book is the most up-to-date summary of the many facets of both the science of our meteorology and how it impacts California's economy and our daily lives. Each chapter has been carefully researched and clearly written. The text is supported by excellent illustrations and graphical materials. All this care in presentation makes this book especially useful for educators and the collections of public libraries. Simply, the authors have created a terrific, broad-ranging, understandable, very readable book on California weather and climate. Released in paperback, the book is a bargain. Treat yourself to a good read about Californians' favorite topic of conversation, "What's happening with the weather?"

William Patzert
Retired Climatologist
NASA's Jet Propulsion Laboratory
Pasadena, USA
https://solarsystem.nasa.gov/people/
1220/william-patzert/

Preface

Most of us in the meteorology profession get hooked by the fascination of the ever changing skies. Steve, having lived in several North American locations and Mike's experience as an Air Force Weather Officer gave us an up-close look at a wide variety of weather and climate types. How the seasons change and influence the yearly progression of hot to cold and rain to snow captivated our interest.

So why write a book on California weather and climate? Some of our friends have given us their books on weather and climate in Wisconsin, Nebraska, and New England. Where was one on California? To find one you have to go back to the 1950s. There have been a lot of new discoveries since then that needed to be shared. El Niño/La Niña, Pacific Decadal Oscillation, atmospheric rivers, and pineapple express are just a few new terms. Then there is global warming, urban heat islands, and the recent megadroughts and increasing wildfires that scream off the front pages of newspapers and over media. California's weather and climate is changing, as it is globally. This book has been calling out to be written for some time and can fill an obvious gap in Earth science literature that focuses on the most populous state.

Our book may be used as a textbook, like for an introductory weather and climate course in college. For non-academics, we wanted to share our love of the atmosphere with readers who also are interested in the diverse climates and highly variable day-to-day weather found from north to south and west to east across a large complex of California's landscapes.

These are exciting times when climate change and global warming are very much in our mindset. In the last decade, the number of disastrous events that have impacted California (and the Earth) has noticeably increased. The hundred year floods and droughts are occurring more than the once a century defined by their names. Frequencies and intensities of these as well as heat waves and fierce wildfires are becoming more common. And there are more people in the state affected by them. This book goes beyond describing the diverse climates of the large state. It intends to contrast the normal with the present changes taking place.

As with other state weather and climate books, we cover the basic elements of weather and climate as pertaining to California. We also include chapters that focus on California's population magnet, its sunshine (Chap. 3). As most Californians live in urban centers, we've included a chapter on air quality (Chap. 12) and a section on urban climates (Chap. 4). Since California is also a leading agricultural state, we add Chap. 13, looking at how California's climates influence productivity in farmlands.

We also include weather and climate information that extends discussions in the chapters found in the Appendices. Appendix A shows the latest climate normals for several California cities, 1991–2020. Appendix B describes some of the major historical winter storms, while Appendix C provides tables and descriptions of the most devastating wildfires in the state's history. Appendix D includes the American Lung Association's State of the Air 2022 report listing the top-ten most polluted cities. And for historical buffs, Appendix E includes excerpts from Richard Dana's (1840) Two Years Before the Mast in which he describes weather conditions in southern California in the 1830s.

We found that the task of writing about the changing weather and climate of California is unending. The conditions affecting the atmosphere continue to evolve. Even as we finished the final drafts of this book in early 2023, the state was suffering from a series of winter storms, floods, and mudslides. Snowfall in the Sierras was headed to new record amounts. And yet, most of the state was still considered in at least moderate drought. We could be adding new material to our chapters indefinitely. But we also are eager to see what our readers think about the present text.

Los Angeles, USA   Stephen LaDochy
Sammamish, USA   Michael Witiw

Acknowledgements

Many individuals provided help and guidance in the writing of this book. The authors would like to thank Jan Hazelton, Sue Thompson and Barb Beck for their help in explaining complex concepts and generally greatly improving the readability of the text. Harrison Blizzard used his skills and developed several original graphics. We also thank Weldon Hiebert for his excellent graphics.

Thanks also to Bill Reid and Paula Arvedson for providing the stunning photos. Thanks go to Bill Selby for proofreading our earlier drafts and making some excellent suggestions. Our writing took a sharp turn after his remarks. Thanks also to Bill Patzert for providing the LA historical temperature chart, writing the Foreword, and for personal inspiration. Kudos to Pedro Ramirez for proofreading as well.

Our sincere thanks goes to Ocean Navigator, Solar Schoolyard, El Dorado, Office of the Washington climatologist, NREL, Direct Relief, PRISM at Oregon State University, American Meteorological Society, Western Region Climate Center, JPL, NASA, Public Policy Institute of California, Jan Null, Mike Pidwirny, Robert Karis and sfog.us, Dave Gomberg and Joe Sirard, NWS FO, Oxnard, Alchetron, Springer Nature, Creative Commons and slideshare.com for permission to use their resources.

Thanks go out to our wives, Sue and Beth, and our families. They put up with many missed weekends and evenings as we spent time away from them working on this book.

We also acknowledge our many friends and co-workers that helped us along the way in our academic and professional careers.

Contents

Acronyms

AL	Aleutian low pressure system
AMS	American Meteorological Society
AR	Atmospheric River
BAMS	Bulletin of the American Meteorological Society
BCE	Before current era (same as B.C.)
CARB	California Air Resources Board
CBD	Central business district
CDD	Cooling degree day
CE	Current era (same as A.D.)
CIMSS	Cooperative Institute for Meteorological Satellite Studies, University of Wisconsin
CPC	Climate Prediction Center, part of NOAA
DRI	Desert Research Institute, part of WRCC
ENSO	El Niño—Southern Oscillation
EPA (US)	Environmental Protection Agency
FAA	Federal Aviation Administration
HDD	Heating degree day
IPCC	Intergovernmental Panel on Climate Change
JISAO	Joint Institute for the Study of the Atmosphere and Ocean, cooperative between the University of Washington and NOAA
JPL	Jet Propulsion Laboratories (NASA facility, under California Institute of Technology)
LA DWP	Los Angeles Department of Water and Power
MJO	Madden-Julian Oscillation
MWR	Monthly Weather Review

NAO	North Atlantic Oscillation
NASA	National Aeronautics and Space Administration
NCEI	National Centers for Environmental Information, part of NOAA
NCEP	National Centers for Environmental Prediction, part of NOAA
NOAA	National Oceanic and Atmospheric Administration
NPH	North Pacific High pressure system
NPO	North Pacific Oscillation
NTSB	National Transportation Safety Board
NWS	National Weather Service
PDO	Pacific Decadal Oscillation
PRISM	Parameter-elevation Regressions on Independent Slopes Model, Oregon State University Climate group
SCAQMD	South Coast Air Quality Management District, part of CARB
SMUD	Sacramento Metropolitan Utility District
TS	Tropical Storm
USDA	United States Department of Agriculture
WRCC	Western Regional Climate Center

List of Figures

List of Tables

1

The Most Climatically Diverse State in the United States

Fig. 1.1 Los Angeles National Weather Service Forecast Office, Oxnard, California. Forecaster Joe Sirard, surrounded by computer screens and Doppler radar screen, is part of a 24/7 days operation

© The Author(s), under exclusive license to Springer Nature Switzerland AG 2023
S. LaDochy and M. Witiw, *Fire and Rain*,
https://doi.org/10.1007/978-3-031-32273-0_1

California is one of the most difficult places in the United States to forecast. It has the greatest diversity of terrain and micro-climates of all the states.—Dave Danielson, NWS forecaster

Movie weatherman, Steve Martin, posted sun, sun, sun on his broadcast for *LA Story*, but the state's weather is a lot more complex. The state is the most diverse topographically, climatically and ecologically (Mitchell, 1976). California is known for both its cultural and geographical diversity. Geographically the differences are striking: California is home to the highest peak in the 48 contiguous United States (Mount Whitney, elevation 14,505 feet) and the lowest spot in the Western Hemisphere (Death Valley, elevation—252 feet). Not only are these two locations located in the same state, but they are actually only 76 miles apart! As a result, California has a climate that is just as diverse as its geography and population. Despite the conventional myth that the Golden State (so-called because of the Gold Rush in the 1850s) is blessed with sunny beach days 365 days of the year, the reality is quite different. Temperature difference between the state's record extremes is actually the fifth greatest in the United States, with a range of 179° between its record minima of—45 °F at Boca Reservoir near Truckee and its sizzling maxima, of 134 °F at aptly named Death Valley, also a North American and world record high (Fig. 1.1).

Myths about the wonderful California climate made it an immigration magnet from early times.

Climate…spell it with a capital, and then try to think of an adjective worthy to precede it. Glorious! Delicious! Incomparable! Paradisiacal!!!" (from an 1893 guide for Southern California traveler)

1.1 California Climates

Over the years, there have been many attempts to classify climates and place similar climates in categories. One still in wide use today was developed by Wladimir Köppen and later modified by Rudolph Geiger in 1961. The Köppen classification has five major climate types and many subtypes. The five major types include tropical (A); dry (B); temperate (C); continental (D); and polar and alpine (E).

California contains all the Köppen climate classifications except tropical (Köppen Type A). Most of the populated regions of California exhibits a Mediterranean climate (Köppen Climate Classification subtype Cs), typified by summers that range from foggy and relatively cool (coastal areas) to hot and dry inland, and mild, wet winters (Fig. 1.2). This type of climate is often referred to as a Mediterranean climate and is similar to what we see adjacent to the Mediterranean from parts of southern Spain and Italy eastward to Greece and Turkey. These areas include the coastal plains, valleys, and coastal mountains from the Oregon border in the north to the Mexican border in the south, along with the Central Valley. Temperatures typically trend from more temperate near the Pacific Ocean to more seasonal extremes toward the state's interior.

The exception to this pattern occurs in California's higher mountains and its deserts. The Sierra Nevada, the Northeast Plateau, and the Mount Shasta region can best be described as a combination of continental climate and a highland climate (Köppen types C and D), while the Mohave and other desert regions in southeast California represent true desert climates

Köppen Climate Types of California

Köppen Climate Type

- BWh (Hot desert)
- BWk (Cold desert)
- BSh (Hot semi-arid)
- BSk (Cold semi-arid)
- Csa (Hot-summer mediterranean)
- Csb (Warm-summer mediterranean)
- Csc (Cold-summer mediterranean)
- Cfb (Oceanic)
- Dsb (Warm-summer mediterranean continental)
- Dsc (Dry-summer subarctic)
- ET (Tundra)

Data sources: 1991-2020 climate normals from PRISM Climate Group, Oregon State University, https://prism.oregonstate.edu; Outline map from US Census Bureau

Fig. 1.2 California's Koppen climates favor mostly Mediterranean types

(Köppen type B). Some of the areas highest in elevation include tundra and glaciated highlands (Köppen type E). Note on Fig. 1.1 that the Cs climate extends from most coastal areas into the Central Valley and even into some mountainous areas.

1.2 Rainfall

Rainfall varies considerably in California. In general, rain totals decrease from north to south but are also highly influenced by the varied topography. For example, much of Marin County just to the north of the Golden Gate receives as much as twice the rainfall of San Francisco. Some mountainous areas in the northwest may reach over 100 inches a year. To the lee of the high Sierra Nevada Mountains, we have the desert areas of southern California where annual rainfall is just over two inches at Death Valley, and 4.61 inches at Palm Springs. These are averages. Year-to-year differences can be just as stark (Fig. 1.3).

As we have seen, like so many other aspects of California, the weather and climate of the state are extremely unique and diverse. There are few other places on Earth where someone can be a couple of hours' drive from the ocean beaches, a desert, or an exciting ski slope. Weatherwise, California truly is the Golden State of diversity. And yet, during the 2020–21 Drought, it was mostly brown.

Unprecedented droughts, worst in a millennium, devastating fires and windstorms, ever-increasing heat waves and record floods are not what most people imagine as sunny California. Yet, the last decade had all of these.

The sunshine state is known for its mild Mediterranean climate, sun-drenched beaches, majestic redwoods, and popular desert resorts. Most outsiders view California as a coastal region with its climate moderated by the cool Pacific, since the majority of residents and tourists remain within 100 km of the coast. They picture the coastal cities along with nearby valley vineyards as the typical California scene. But that vision only covers the "perceived" California, a small portion of a large, highly diverse state. On many summer days, the hottest and coldest spots in the lower 48 states occur in California, such as at Death Valley (Fig. 1.4) and Bodie State Park, with summer temperatures reaching into the 120's and also 20's °F, respectively. Of course, the former is located at −86 m (−282 ft.) below sea level, while the latter resides in the Sierras at 8375 feet or 2553 m.

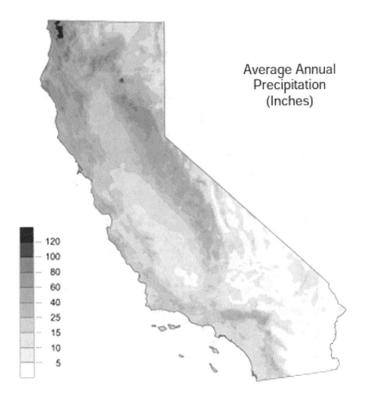

Average Annual
Precipitation
(Inches)

120
100
80
60
40
25
15
10
5

Fig. 1.3 California annual rainfall totals (Kauffman, E. Atlas of the Biodiversity of California—Climate and Topography)

California weather is not always mild. In the winter of 1861–62, large portions of the Central Valley and coastal southern California were large lakes, a result of phenomenal rainstorms that persisted through December and January. The worst flooding in California history occurred with 28 days of rain between December 24, 1861 and January 21, 1862, was the equivalent of at least a 30,000-year event. Record, or near record flooding extended from Humboldt and Eureka in the north to Orange and San Diego counties in the south, with the Central Valley under water, while floods washed out much of the southern California coastal plains.

More recently, from January 3 to 5, 1982, torrential rains caused extensive damage and destruction in the lowlands of the central and northern parts of California and heavy snows fell in the highest mountains. The Santa Cruz Mountains were inundated with 10 to 20 inches of rain in 30 h. The National Weather Service reported more than 8 inches of rain in one day, the greatest 24-h rainfall since 1890 when record keeping began. Considered one of the worst storms of the century, several thousand people were flooded out of their

Fig. 1.4 Badwater Basin in Death Valley NP is the hottest place on Earth. (National Park Service, Death Valley National Park)

homes, at least 18 people killed, trains derailed, and schools and highways closed. And they say California has no weather?

More often it's the lack of rain that plagues the state. Megadroughts of over 40 years have been detected from proxy data such as tree rings. The longest dry spell in the United States took place at Bagdad, California, where it didn't rain from October 3, 1912, until November 9, 1914, a period of 767 days. The 2020–22 Drought may stand as one of the worst in state history.

Heat waves, fires, and Santa Ana winds often go together to present another detour from mild California weather. Hot, dry winds in fall or winter can send temperatures soaring above 100 °F and/or whip small brush fires into gigantic inferno wildfires. The unusually long heat wave of 2006 (July 16–26) resulted in 163 fatalities and large agricultural losses, including death of more than 25,000 cattle and 700,000 fowl. Heat waves have become more frequent and longer in duration, especially in the southern California region. While Santa Ana winds have not appreciable increased, their impacts through wildfires have increased in both northern and southern California. In October 2007, nearly 250,000 acres of southern California burned in the Witch's Fire, costing $1.3 billion, and causing the largest evacuation in state history, 1 million were evacuated. While in August 26, 2009, the Station Fire in Los Angeles was the largest fire in the LA County history (Fig. 1.5). Oakland Fire (referred to as the Tunnel Fire) was another destructive fire in October 19, 1991 driven by Diablo winds of 60 mph. Originating in the steep hillside above Berkeley near the Caldecott Tunnel, fire ultimately killed 25 people and injured 150 others. Santa Ana winds, even without accompanying fires, can be tremendously destructive. These warm, dry winds can accelerate through narrow canyon passes reaching hurricane strength.

The mild, Mediterranean climate also includes El Niño and La Niña events that can bring destructive weather to the state. Two of the largest El Niño events took place in the winters of 1982–83 and 1997–98. Both brought unusually high amounts of rain and snow as well as coastal erosion. More on El Niño/La Niña is in Chap. 5.

While not known for its cold winters, the Golden State is known for some of the highest snowfall in the US. The western slopes of the Sierra Nevada capture tremendous amounts of white stuff measured in tens of feet. Heavy snowfall can be destructive and sometimes fatal, as with the infamous Donner

Fig. 1.5 Station Fire in September 2009 threatened NASA/Jet Propulsion Laboratory below in Pasadena caused by arson, it was the largest fire in Los Angeles County history—NASA

Party in the winter of 1846–47. And yet, Sierra snowfall is the main source of the state's water supply. Snowfall will be discussed in Chap. 6.

Tropical storms, which are common on the east coast, rarely impact even the southernmost portions of the state. Eastern Pacific hurricanes normally form in warm waters off Mexico and Central America. Since they require warm (at least 26 °C, nearly 80 °F) water to sustain their energy, most Pacific tropical storms dissipate as they cross into the cooler water of coastal Baja California. Sometimes, the remnants of weakening storms are carried by winds into southern CA. Mostly clouds and possibly scattered showers may result within the state. In September 1938, a weakening tropical storm reached southern CA, causing flooding rains, over 11″ in mountainous Tehachapis of the Traverse Range, washing away the Southern Pacific railroad. A powerful Hurricane Linda also tracked into the SE deserts producing scattered showers and a reported tornado in southeast California in 1998. During winter, the Pineapple Express (Hawaiian atmospheric river) may bring tropical moisture into coastal California energizing polar cyclones and causing heavy rainfall. In December 2010, a river of atmospheric moisture carried unusually moist air into southern California setting record rainfall from a six-day series of storms.

Diversity and variability are more the norm than the normal. Like other regions of the US, there are extremes in heat and moisture in the Golden State. Recent weather records have fallen like dominos. The misconception that California does not have weather or seasons has been exposed, especially in 2020 and 2021 with record heat, drought, and fires.

Bibliography

California Biomes. (2022, December 2). https://civilizedape.weebly.com/biomes-of-california.html

California Chapparal Institute (2022). https://www.californiachaparral.org/threats/too-much-fire/

California Soil Resource Lab, University of California Davis. (2019, December 2). *GIS and digital soil survey projects.* https://casoilresource.lawr.ucdavis.edu/projects/pedology-and-soil-survey/gis-and-digital-soil-survey-projects/

Kauffman, E. (n.d.). *Atlas of the Biodiversity of California—Climate and topography.* California Coastal Voices. https://www.coastal.ca.gov/coastalvoices/resources/Biodiversity_Atlas_Climate_and_Topography.pdf

Mitchell, J. M. (1976). An overview of climatic variability and its casual mechanisms. *Quaternary Research, 6,* 481–493.

National Park Service. (2022, December 1). *NPS Geodiversity Atlas—Index.* https:/
/www.nps.gov/articles/geodiversity-atlas-map.htm
Peterson, A. (2022, December 2). *Köppen types calculated from data from PRISM
Climate Group,* Oregon State University. http://prism.oregonstate.edu

2

The Golden State and Its Physical Environment

Fig. 2.1 Shaded relief map of California shows the diverse and rugged topography. (The National Atlas: United States Geological Survey)

California is the third largest state after Alaska and Texas. Its area includes nearly 163,000 square miles extending from 42° north latitude at the Oregon border to just north of 32° north latitude at the southwest corner of the state along the Mexican border. On the east, the state borders Nevada following the 120° west longitude line from Oregon to Lake Tahoe. The border follows

© The Author(s), under exclusive license to Springer Nature
Switzerland AG 2023
S. LaDochy and M. Witiw, *Fire and Rain*,
https://doi.org/10.1007/978-3-031-32273-0_2

another straight line southeast to where shared Nevada and Arizona borders at the Colorado River. It then follows the River south until it reaches the Mexican border. On the west, the coastline extends from near 117° west longitude at the Mexican boundary, northwest to the Oregon border. The northwest–southeast tending state stretches about 9½° of latitude and over 10° of longitude.

2.1 Topography and Its Role in Weather and Climate: Physiographic Regions

Much of the state is mountainous (Fig. 2.1). Geologic internal forces associated with volcanic and seismic (earthquakes) activities are responsible for these mountains. Mountainous terrain, aligned mostly in a north–south direction, is a major factor in local and regional temperature, precipitation, and wind patterns. Mountains also determine the natural distribution of plants and animals, soil types, and water drainage. They also play a role in location of population centers. Figure 2.2 shows the physiographic regions which will be briefly described.

2.1.1 Klamath Mountains, Northwest

Here, steep, moist, and heavily forested mountains are circled by the Pacific Ocean on the west, Oregon to the north, Shasta Valley and the Cascades to the east, and the Sacramento Valley on the southeast. Winter storms make this region one of the wettest, with some areas receiving over 100 inches of annual precipitation on westward facing slopes. There is a large gradient of climatic zones away from the coast vertically from hot summers and mild winters in the river valleys to cool summers and frigid winter near mountain summits.

2.1.2 Southern Cascade Range

Found east of the Klamath Mountains, this north–south oriented area includes the southernmost portion of the Cascades with its two dominant volcanoes, Lassen (elevation 10,457 feet.) and Shasta (14,162 feet). While the Cascade valleys are drier than valleys facing the Pacific, precipitation increases with elevation. Mount Shasta receives plentiful snow in winter, making it a favored ski area. Being farther inland, this climatic region is more continental

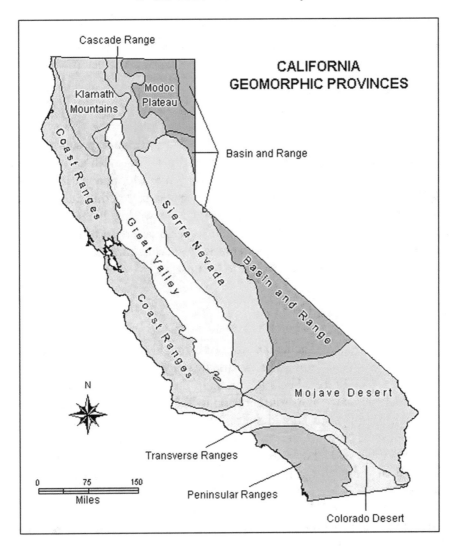

Fig. 2.2 California's geomorphic provinces or physiographic regions. (California State University Northridge: Map Library)

than the Klamath Mountains. Like the Klamath region, there are multitude of microclimates due to elevation, slope, and orientation.

2.1.3 Northeast, Modoc Plateau

The northeast plateau region is dominated by volcanic landforms, consisting of hard, black basaltic rock from lava flows, similar to the Columbian Plateau to the north. Being in the rain shadow of the Cascades and Klamaths, the region is relatively dry (about 20 inches of precipitation per year). The winters are longer and colder here, with more extreme seasonal temperatures.

2.1.4 Basin and Range Province

Much of this region lies to the east of the Sierras extending south to the Garlock Fault and the Mohave. The ranges run north–south and are quite high, including the White Mountains, with White Mountain. Peak at over 14,000 feet. These high ranges border elongated basins, including Death Valley, the hottest spot on the continent (and the world). Being in the rain shadow of the Sierra Nevada, the region's precipitation and vegetation are sparse even at higher elevations, where the bristlecone pine, the oldest living trees in the world, stands in the White Mountains.

2.1.5 Sierra Nevada

This granitic barrier to Pacific moisture runs roughly parallel to the coastline in a north–south direction. The mountain range is 40–60 miles wide and rises to over 14,000 feet. The High Sierra in the central portion of the range includes Mt. Whitney at its southern end, which at 14,505 feet is the highest point in the contiguous United States. Nearby, at only 76 miles away, is the lowest point in North America, Death Valley, at 282 feet below sea level. At the northern end, the Sierra Nevada joins the Southern Cascades, while the southern end of this geomorphic province joins the Transverse and Coast Ranges.

In Spanish, Sierra Nevada translates into snow covered mountains. Its name is very appropriate as this range has some of the snowiest places in the world and holds records for snow totals in the United States. This huge barrier to air flow provides abundant precipitation on the windward (western) side and a rain shadow effect on the dry leeward (eastern) side. Climatic zones follow the changes in elevation from temperate to frost climates on the western side and desert conditions on the lower eastern side. More on the Sierra Nevada snows in Chap. 6.

2.1.6 Central Valley

The Central (or Great) Valley lies between the Coastal Mountain Ranges and the Sierra Nevada. It extends 400 miles from the Klamath Mountains to the north, near Red Butte, to the Tehachapi Mountains on the southern end near the city of Bakersfield. The valley is roughly 50 miles wide on average, with about 25,000 square miles of productive agricultural land due to the depression being filled with sediments from the surrounding mountains. Many active streams contribute to the rich soil with two major streams dividing the valley into northern and southern sectors. The Sacramento River Basin in the northern portion covers about one-third of the valley, while the San Joaquin River Basin covers most of the remaining two-thirds on the southern end. The extreme southern portion is the Tulare Basin with its semi-arid climate. The Sacramento River flowing south meets the San Joaquin flowing north in the delta area of San Francisco Bay spilling into the Pacific Ocean.

The Sacramento River carries considerably more water than the San Joaquin. The latter is reduced by intensive agricultural and urban water use, especially with the recent drought. Water demands throughout much of the state have intensified for the precious water coming from these rivers.

Summers can be blistering hot throughout the Central Valley, while winters are cool and damp, gradually becoming drier toward the south. The Central Valley's climate is classified as "interior Mediterranean." The southwest portion of the San Joaquin Valley, near Bakersfield, is a subtropical desert, not quite as hot as the Colorado Desert, with winters milder than the Mohave. Fog, known as "Tule fog" regionally, frequently reduces visibility in winter. North Pacific storms become much drier than along the coastal mountains, with annual rainfall ranging from 23 inches at Red Bluff, and 19 inches at Sacramento, to barely seven inches near Bakersfield. The west side of the valley is in a rain shadow and has less precipitation than the east side. Fortunately, the main watershed for the Central Valley is the Sierra Nevada, where snowpack provides a large natural reservoir. That reservoir does vary widely in the quantity of water it provides from year to year.

2.1.7 Southern California Deserts: Mohave and Colorado

To the east of the Sierra Nevada, the Basin and Range geological province is in an extreme rain shadow region. The dryness continues in the Mohave Desert. Cutoff from most Pacific moisture, these desert regions experience hot summers and chilly winters. The Mohave Desert is situated north and east of the Transverse Ranges. It typically has summer temperatures above 100 °F. Like many continental deserts, the Mohave can be quite cold in winter with strong winds. Temperatures vary greatly with elevation, cooling in the higher mountains. Annual precipitation ranges from 3.5 inches at lower elevations to nearly 10 inches in the mountains. Occasional snow falls in the mountains. East of the Peninsular Ranges, the Colorado Desert tilts lower toward the Mexican border and includes the man-made Salton Sea. Like the Mohave, temperatures can be quite extreme. Summer thunderstorms (associated with the North American monsoon) may bring sudden, heavy rainfall to both desert regions. The driest months in the southeast deserts are May and June, before the monsoon season (see Chap. 5 for monsoon rains).

2.1.8 Southern California Coastline and Coastal Mountains

In the southwest part of the state, where the majority of the state's population resides are the Peninsular Ranges. With a north–south orientation, these coastal mountains extend from the Transverse Range to the north into Baja California to the south. These parallel Peninsular Ranges create hundreds of microclimates, as sea breezes, coastal stratus, and Pacific storms have their greatest influence on the windward side of these ranges, while leeward sides are relatively dry and more continental. The seasonal changes in the coastal marine layer provide the milder conditions to cities along the Pacific.

2.1.9 Transverse Ranges

Unlike the Coast Ranges and the Sierras, the Transverse Ranges run west–east, the only major mountain range in the state to do so. This twisted sister range got turned around as the two tectonic plates, the North American and the Pacific, spun their ends clockwise in opposite directions. Major mountains in this region include, from west to east, the Santa Inez Mountains, Santa Susana Mountains, Santa Monica Mountains, San Gabriel Mountains, and the San Bernardino Mountains. Included in this province are the lowland Oxnard Plain, the San Fernando Valley, and the portions of the Channel Islands. The Transverse Range also forms the southern boundary for the Central Valley to the north. Winter travel over the range can be hazardous as ice, snow and fog can disrupt or shut down traffic. Winter storms rising from the southwest can produce some of the highest rain amounts in the state.

2.1.9.1 Central Coast and Coast Ranges

The Coast Ranges are northwest-trending mountain ranges (2000 to 4000, occasionally 6000 feet elevation above sea level), and valleys to the north of the Transverse Ranges. The ranges and valleys run parallel to the San Andreas Fault with one major break at San Francisco Bay. To the west is the Pacific Ocean which has direct influence on this province. The cool water tends to keep the coast quite cool throughout the year, while further inland its influence becomes less. Summer low clouds and fog hug the coastline and may extend inland where there are gaps in coastal uplands. These ranges cause drier, sunnier conditions on the lee of coastal terrain so that temperatures become more extreme with inland coastal valleys. Temperature gradients across the San Francisco and Los Angeles metropolitan areas can be quite large on summer days, being cool near the coast and hot just a few miles inland. Precipitation increases to the north in winter and can be heavy in steep topography as in the mountains inland of Santa Cruz. In general, rainfall increases with elevation.

2.2   Vegetation and Agriculture

The varied topography is one reason California ranks as the most biologically diverse state. The state also is home to species that do not exist anywhere else, such as the sequoia trees (*Sequoiadendron giganteum*) found only on the western slopes of the Sierras. The vegetation varies in the many microclimates found throughout the large state. These unique florae include the northern coastal redwoods, the Mohave Joshua trees, the sequoias in the Sierras, the largest plants in the world, and the chaparral of the coastal hillsides. Figure 2.3 shows the distribution of the primary biomes in the state. A major contributor to the vast diversity of plant species is elevation. For example, a cross section over the central Sierra Nevada shows rapid cooling and increased precipitation with increasing height on the western slopes, while a marked rain shadow and dry climate species dominate on the eastern slopes. In the far northwest, redwoods and Douglas fir occupy the coastal strip and lower river valleys of the region. At mid-elevations, there is a mixed forest of evergreens, while farther inland are more drought-adapted pines and oaks.

It is not surprising that California, with its varied range of climates and topography, also has the most diversity of plant species on the non-tropical portion of the continent. The rich variety of trees, evergreen and deciduous, grasses, and shrubs are found in abundance throughout the state.

Natural resources, such as excellent flat fertile valleys, the moderating effect of the cool California Current, a long growing season, the mild, Mediterranean climate near the Pacific coast, have helped California become the

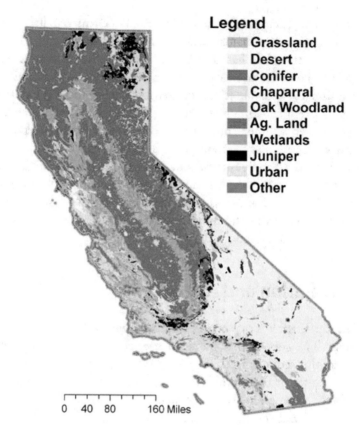

Fig. 2.3 Major biomes of California. (University of California Davis: California Soils Resources Lab)

leading agricultural state in the United State. See Chap. 13 for agriculture and weather.

2.3 California Water Resources

During most years, California has sufficient water supplies for all its needs (agriculture, urban, industry, and environment). In wet years, unused water is either stored in reservoirs or sent back to the ocean. In dry years, the amounts of water may not be adequate to meet the needs of all users. Rainfall and snow are more plentiful in the northern regions (2/3 of the total supply), while demand is greater in the drier southern regions. To fix this imbalance, the state created one of the most elaborate water transfer systems in the world (Fig. 2.4). The California and Los Angeles Aqueducts bring Sierra

Nevada water south, while the Colorado River Aqueduct taps western mountains' snowmelt for thirsty southern California users. When all three of these sources run dry, as in 2020–22, you have what UCLA Geographer, Glen MacDonald, calls *The Perfect Drought*.

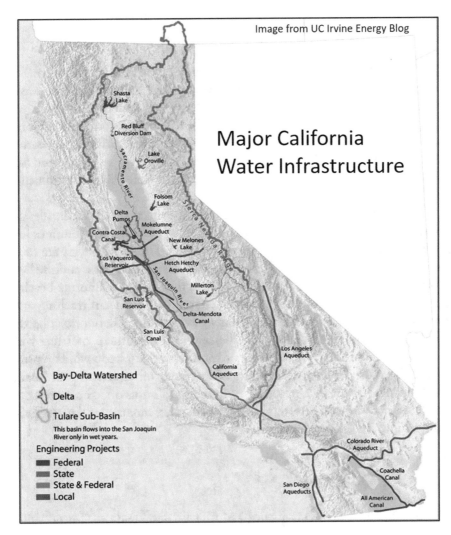

Fig. 2.4 State's water distribution system carries water from where it is most plentiful to where the demand is greatest (University of California, Irvine Energy Blog 2015)

Groundwater is also an important source of moisture, especially in the drier portions of the Central Valley for irrigation. Decades of irresponsible pumping from falling water tables led to the Sustainable Groundwater Management Act, 2014, regulating how much withdrawals could be made.

2.4   Ocean Influences

The Pacific Ocean forms the western boundary of the state but also is one of the most important influences to California's weather and climate. The ocean moderates air temperatures and is the source of most water vapor that eventually falls as rain and snow. The majority of precipitation comes from North Pacific storms, which will be discussed later, while sea surface temperature patterns such as the El Niño/La Niña and the Pacific Decadal Oscillation (see Chap. 6) control climate variability throughout the state. They also influence much of the marine life along the coast.

The cold California Current also chills the surface air above during the warm season, producing cool ocean breezes along the coast and summer low stratus and fog, especially along the central coast. The ocean effects are most prominent in summer, when the contrast between the cool coast and the heat of the interior valleys are the greatest visitors to southern California beaches on a hot summer day are surprised by the refreshingly cool surf reaching only near 68 °F by the end of summer. May Gray and June Gloom describe the cool, cloudy days that often shroud the central and southern coastline with low stratus clouds and fog, a result of warming atmospheric winds passing over chilly coastal waters. These low layer clouds often extend well inland until the intense sunshine of longer summer days "burns off" the overcast by late morning. May and June tend to be the cloudiest months for the otherwise sunny, southern California coastline.

2.5   Air Quality

The state has diversity in air quality. Los Angeles Basin and the Central Valley have some of the worst air pollution in the nation. During recent northern California fires, San Francisco and adjacent regions had some of the highest particulate levels in the world! On the other hand, remote, rural areas of the state also have some of the cleanest air in the United States. Air pollution and weather are discussed in Chap. 12.

Bibliography

Anderson, B. R. (1975). *Weather in the West*. American West Publishing Company.

California Soil Resource Lab, University of California Davis. (2019, December 2). *GIS and digital soil survey projects*. https://casoilresource.lawr.ucdavis.edu/projects/pedology-and-soil-survey/gis-and-digital-soil-survey-projects/

Hyslop, R., Lin, W., & Garver, S. A. (2009). *California eclectic: A topical geography*. Kendall/Hunt.

Rutten, T. (2008, October 15). Fire, the price we pay. *Los Angeles Times*. https://www.latimes.com/archives/la-xpm-2008-oct-15-oe-rutten15-story.html

Silverman, D. (2022, December 4). *California water projects feeding Southern California*. University of California, Irvine Energy Blog. https://sites.uci.edu/energyobserver/2015/04/28/california-water-projects-feeding-southern-california/

MacDonald, G., Kremenetski, K. V., & Hidalgo H. G (2008). Southern California and the perfect drought: Simultaneous prolonged drought in southern California and the Sacramento and Colorado River systems. https://www.sciencedirect.com/science/article/abs/pii/S1040618207001966

UC Irvine Energy Blog. (2015, April 15). *California water projects feeding southern California*.

3

California Sunshine

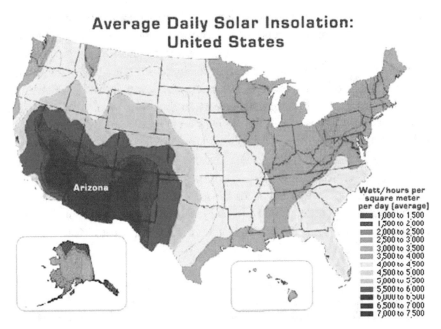

Fig. 3.1 Average daily solar insolation (watt/hrs/sq m/day) for the United States (NOAA)

"Let the sunshine, let the sunshine in.." Aquarius-Ragni, Rado and McDermot from the musical "Hair."

© The Author(s), under exclusive license to Springer Nature
Switzerland AG 2023
S. LaDochy and M. Witiw, *Fire and Rain*,
https://doi.org/10.1007/978-3-031-32273-0_3

3.1   Solar Radiation

Contrary to public opinion and several pop songs, California is not always sunny. Ask Tony Bennett when he sings about leaving his heart in San Francisco. Late spring and early summer along the southern and central California coast can be socked in with low stratus clouds and fog (see Chap. 7). Winter Tule (radiation) fog can also block sunshine in the Central Valley for weeks at a time. However, it is true that the state gets its fair share of sunny days. Much of the southeastern deserts receive the highest annual average solar radiation in the lower 48 states (Fig. 3.1). In terms of average annual solar radiation, the amounts decrease from the southeast of the state toward the northwest. The sunniest place in the lower 48 states is in Imperial County, California. It's a great place to grow crops, if you have water, from the Colorado River.

Solar insolation, or the amount of solar radiation that reaches the ground, determines how much energy is available. This radiation is needed for photosynthesis in plants and for converting the sun's energy into heat. Seasonal differences can be quite large as latitude determines both length of daylight and the intensity of insolation. Summer daylight is longer, and solar altitude is also higher, producing more direct energy at lower latitudes. Figure 3.2 shows the sun path for different seasons at San Francisco. The noon sun varies from 28.5 degrees above the horizon on the first day of winter to 75.5° on the first day of summer, while the length of daylight increases from 9 h, 33 min to 14 h, 47 min. But surface features also determine heating potential. Surface color, texture, slope, and composition determine how much heat is retained and how much is lost to the environment. In cities, urban materials and building geometry are responsible for the urban heat island, making for warmer conditions than in rural or vegetated landscape. More on urban heating is in Chap. 4.

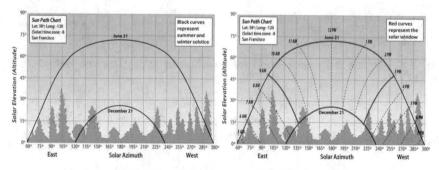

Fig. 3.2 Sun path for San Francisco at different seasons. Red curve shows times when trees do not block sunlight. (Clay Atchison, www.solarschoolhouse.org)

3.2 Sunshine

While insolation is measured with radiometers, sunshine is measured by weather personnel by taking observations of the sky. Each hour, an observation is taken of how many tenths of the sky are covered with clouds. Such weather data is collected by the National Centers for Environmental Information (NCEI) which can be tabulated into the number of days with sunny, partly sunny (or partly cloudy) or cloudy skies. Table 3.1 shows the average annual number of days with sunny and partly sunny skies for cities across the state.

In the table, the average number of sunny days for a city in California is the total days in a year when the sky is mostly clear and has at least 70% of the available sunshine. On partly sunny days from 40% to less than 70% of the available sunshine is reported. Total days with sun are a sum of sunny days and partly sunny days. Table 3.1 shows annual averages, calculated from decades of weather observations.

California certainly has its share of sunny days. During July, Sacramento is one of the sunniest places on earth with 98% of the available sunshine. Hours of daylight increase in summer, peaking at Summer Solstice (June 21 or 22). Summer daylight also increases with latitude. For instance, in San

Table 3.1 Annual average days of sunny and partly sunny days for various California cities

Annual days of sunshine

City	Sunny days	Partly sunny days	Total days with Sun
Bakersfield	191	81	272
Bishop	201	89	290
Blue canyon	174	64	238
Eureka	77	102	179
Fresno	194	73	267
Long Beach	159	119	278
Los Angeles Airport	147	116	263
Los Angeles Downtown	186	106	292
Mount Shasta	164	83	247
Redding	172	77	249
Sacramento	188	77	265
San Diego	146	117	263
San Francisco Airport	160	100	260
Santa Maria	176	110	286
Stockton	184	77	261

Current Results. Data from NEIC, NOAA

Diego, near the Mexican border (32° north latitude), the longest daylight hours are 14 h, 18 min, while at the Oregon border (40° north latitude) it increases to 15 h, 16 min. During the winter solstice, higher latitudes have shorter hours of daylight by again an hour, from nearly 10 h in San Diego to 9 h, 5 min, near the Oregon's border.

3.2.1 Seasonal Sunshine Hours for Selected Airports

Percentage of possible Sunshine.

Table 3.2 shows that San Francisco Airport (SFO) has the highest percentage of sunshine in spring and secondly in fall, with the lowest during the rainy season, October through February. Summer fog and low clouds are common, especially in July and August though conditions often improve by afternoon. San Diego and Los Angeles along the south coast show a minimum of sunshine during the late spring, early summer fog season (May Gray, June Gloom). By late June, unlike San Francisco, fog is less common as ocean temperatures and land temperatures more frequently keep air temperatures above the dew point.

Interestingly, SFO has as many sunny days as the airports at Los Angeles and San Diego in April, but more than both of these airports in May and June when fog and low clouds are more prevalent along the southern California coast (Table 3.1). This is likely due to the fact that SFO is protected from fog and, to a certain extent, low clouds during the summer months. Many a July day will be sunny at SFO, but foggy or overcast in much of the city of San Francisco.

As the winter rainy season is shorter with both lower amounts and frequency, winter sunshine increases traveling south in the state. Sacramento, like much of the Central Valley, is foggiest in December and January with Tule fog, a type of radiation fog (see Chap. 7). Temperatures are cool, below 50 °F as overcast skies may linger several days, or even weeks. The valley heats up during the summer with some of the highest numbers of cloudless

Table 3.2 Monthly sunshine percentages for selected California cities

% possible sunshine	J	F	M	A	M	J	J	A	S	O
San Francisco	61	69	73	78	74	70	70	68	73	71
San Diego	72	71	70	68	59	58	68	70	69	68
Sacramento	48	67	76	82	92	95	98	94	92	85
Los Angeles	71	72	72	78	64	64	83	84	75	73

Source NCEI, NOAA

days. Although Yuma, Arizona is the sunniest city in the United States when looking at the year-round numbers, Sacramento has the highest percentage of sunshine during the summer months Two other California cities made the top-ten sunniest cities in the United States—Redding (88%) and Fresno (79%). Los Angeles, San Diego and San Francisco, represented by their airports, trail with 73%, 68%, and 66% annual sunshine, respectively.

On average, 354 days in Palm Springs each year are sunny, and the percentage of available sunshine is about 85%; not quite a record, as places a little further east such as Yuma in Arizona have even less clouds, and up to 90% sun. In Orange County, south of Los Angeles, Irvine gets the great weather (it's sunny 73% of the time). There are sunnier spots on the list: Orange County neighbor Santa Ana and Disneyland's home of Anaheim are both 80% sunny. Not typically thought of as sunny California, Eureka on the northwest coast manages only 51% of the available sunshine.

3.3 Solar Radiation Measurements

Aguado(1986), from San Diego State University, collected a year of radiation data for stations in western San Diego County. He found that coastal stations had lower summer radiation values than inland due to coastal advection fogs. The pattern is reversed in winter when effects from the varying terrain are prominent inland. Daily variability between stations was much greater. The decreases in radiation from April to May for all stations reflect the persistent low clouds and fog.

3.4 Solar Energy

Comfortably ahead of its rivals, California remains the undisputed leader when it comes to solar power in the United States, with almost 25 GW of installed solar energy. Solar electricity production in California has two main types, solar thermal and solar photovoltaic (PV) (Fig. 3.3).

Most solar thermal facilities are found in the desert southeast of the state, the Mohave. Solar thermal plants heat water into steam that turns turbines to create electricity. Solar photovoltaic systems convert sunlight into electricity directly. In 2020, solar PV and solar thermal power plants produced 48 gigawatt hours (GWh) of energy or 22 percent of the state's total electricity production and employed more than 74,000 in the sector. Solar PV systems are now producing more electricity than solar thermal plants. A majority of

Fig. 3.3 Solar panels on the Engineering Building, California State University, Los Angeles, facing south, while center panels rotate to follow the sun (Photo by S. LaDochy)

new projects are customer-owned residential PV systems. California's goal is to increase renewables (wind and solar). Electric utility companies were required to use renewable energy to produce 33 percent of their power by 2020, which the state achieved. A main source of renewable power is solar energy.

Solar energy capacity grew tenfold in the last five years (Fig. 3.4). The leading counties in solar energy production are Kern, Riverside, and San Bernardino, in that order. The top city in the nation for total PV energy generated is Los Angeles at nearly 5500 megawatts (MW) in 2020. Total small building rooftop solar potential is 5444 MW. San Diego is second in the nation producing over 420 MW, with a potential of 2200 MW for small building roofs. The state is approaching one million solar systems on rooftops.

Despite its weather, San Francisco is still among the top-ten cities in California with the most solar panels per household (1923 residential rooftop solar installations). Solar panels in San Francisco may generate enough energy over their 20–25-year life spans to repay homeowners for their investments. However, desert communities, like Apple Valley, where there is much more solar energy than San Francisco, according to the National Renewable Energy Laboratory, only one in 90 homes use solar. San Diego, on the other hand, has 2238 residential solar installations, the most in the state. And Central

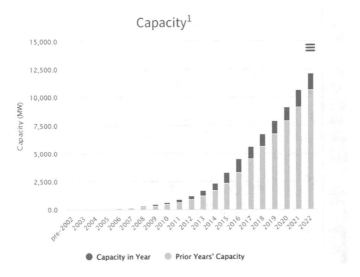

Fig. 3.4 Solar energy capacity continues to increase in the state (CA Energy Commission)

Valley communities, with more sun, like Fresno, Clovis, and Bakersfield, also rank among the state's top-ten solar cities.

Despite an area of over 2200 square miles, Los Angeles County has been able to produce a solar map for the county. It allows you to click on any point on the map to determine solar potential. San Francisco has also created its own solar power map (Fig. 3.5). At San Francisco's Moscone Convention Center, solar panels have produced an average of over 1600 kW hours per square meter per year (kWh/m²/year), despite panels being laid flat and receiving only indirect sunlight.

Solar served up an unprecedented 50% of the state's electricity demand on a sunny day around 1 pm PST on March 5, 2018. The next day, utility operators reported a second record for total generation from solar which produced 10,411 MW, beating out the previous year's record by %. The state is regularly sending excess electricity to Arizona and other states (sometimes paying them to accept) to avoid overloading its own power lines.

On May 9, 2019, California gave the solar industry a huge boost by requiring solar panels on most new homes built after January 2020. The historic measure makes the Golden State the first to require solar installations on most single-family homes, as well as multi-family residential buildings as high three stories, including condos and apartment buildings.

One thing could tip California's solar market back into a furious growth period: a new target. With the mandate to generate half the state's electricity

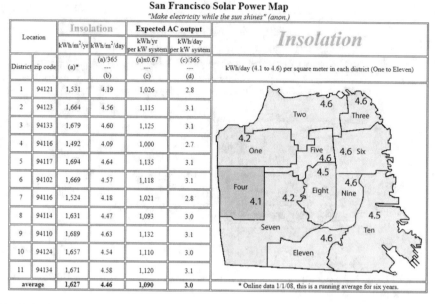

San Francisco Solar Power Map
"Make electricity while the sun shines" (anon.)

Location		Insolation		Expected AC output		Insolation
		kWh/m²/yr	kWh/m²/day	kWh/yr per kW system	kWh/day per kW system	
District	zip code	(a)*	(a)/365 --- (b)	(a)x0.67 --- (c)	(c)/365 --- (d)	kWh/day (4.1 to 4.6) per square meter in each district (One to Eleven)
1	94121	1,531	4.19	1,026	2.8	
2	94123	1,664	4.56	1,115	3.1	
3	94133	1,679	4.60	1,125	3.1	
4	94116	1,492	4.09	1,000	2.7	
5	94117	1,694	4.64	1,135	3.1	
6	94102	1,669	4.57	1,118	3.1	
7	94116	1,524	4.18	1,021	2.8	
8	94114	1,631	4.47	1,093	3.0	
9	94110	1,689	4.63	1,132	3.1	
10	94124	1,657	4.54	1,110	3.0	
11	94134	1,671	4.58	1,120	3.1	
average		1,627	4.46	1,090	3.0	* Online data 1/1/08, this is a running average for six years.

Fig. 3.5 San Francisco Solar potential map (Map courtesy of Robert Karis, www.sfo g.us)

from renewable sources by 2030 now easily within reach, state proposes to require 100% of the state's power come from renewable sources by 2045.

Bibliography

Aguado, E. (1986). Local-scale variability of daily solar radiation-San Diego County, California. *Journal of Climate and Applied Meteorology, 25*, 672–678.

California Energy Commission. (2021). *Developing renewable energy.* https://www.energy.ca.gov/about/core-responsibility-fact-sheets/developing-renewable-energy

ESRI. (2012). California Sunshine is an energy goldmine. *Arc News*, Winter 2011/12. http://www.esri.com/news/arcnews/winter1112articles/california-sunshine-is-an-energy-goldmine.html

Freakonomics. (2011). *Why California's push for solar is a foggy idea.* https://freakonomics.com/2011/08/the-inefficiency-of-californias-push-for-rooftop-solar/

Hernandez, R. R., Hoffacker, M. K., Murphy-Mariscal, M. L., & Allen, M. F. (2015). Solar energy development impacts on land cover change and protected areas. *Proceedings of the National Academy of Sciences, 112*, 13579–13584.

Ingraham, C. (2015). Map: Where the sunniest and least sunniest places are. *Washington Post.* https://www.washingtonpost.com/news/wonk/wp/2015/07/13/map-where-americas-sunniest-and-least-sunny-places-are/?noredirect=on&utm_term=.c66dc6d73c12

McGough, M. (2017). Sacramento's summers aren't the hottest on earth-but could they be the sunniest? *Sacramento Bee*, June 24 online. https://www.sacbee.com/news/weather/article158007444.html

National Center for Environmental Information. (2022). *Comparative Climatic Data (CCD)*. https://www.ncei.noaa.gov/products/land-based-station/comparative-climatic-data

Roth, S. (2020). California proposes big changes to the rooftop solar industry. *Los Angeles Times*, December 13, 2021. https://www.latimes.com/business/story/2021-12-13/california-proposes-big-changes-to-rooftop-solar-incentives

SFOG.us. (2022, December4). *San Francisco solar power map*. https://sfog.us/solar/sfsolar.htm

Solar Map Application. (2022). https://apps.gis.lacounty.gov/solar/m/?viewer=solarmap

State of California, Department of Water Resources. (1978). *California Sunshine-solar radiation data*. Bulletin 187, Sacramento, California.

Wikipedia. (2021). *List of cities by sunshine duration*. https://en.wikipedia.org/wiki/List_of_cities_by_sunshine_duration

U.S. Energy Information Administration. (2021). *Solar explained: Where solar is found and used*. https://www.eia.gov/energyexplained/solar/where-solar-is-found.php

4

Temperature Variations: Heat and Cold

Fig. 4.1 Death Valley is the hottest place on Earth (courtesy of Bill Reid)

Shortly after moving to San Jose, one of the authors left his home on a sunny August day and drove north to a Giants game in San Francisco. Enroute, he was surprised to hear the following weather report: San Jose, sunny and 80; Santa Rosa, sunny and 85; Sacramento, sunny and 100; Oakland, partly cloudy and 59; and San Francisco, overcast and 54. Arriving at the ballpark,

S. LaDochy and M. Witiw, *Fire and Rain*, https://doi.org/10.1007/978-3-031-32273-0_4

the first thing he noticed was the sign that said: "HEATERS ARE NOT GUARANTEED TO WORK." At the concession stands, hot chocolate was selling faster than beer. After shivering through the game, he returned to the warmth of San Jose. This brought to mind the perceptive words attributed to Mark Twain: "The coldest winter I ever spent was a summer in San Francisco." Although the summer climate of San Francisco has warmed a few degrees since the 1970s (due to factors we will examine shortly), the temperatures he experienced would have been typical of a summer day in the 1960s and 1970s (Fig. 4.1).

Death Valley, July 13, 2021, had the hottest temperature ever verified on Earth, 130 °F tying the record set the year before on August 16, 2020. Though an even hotter temperature of 134 °F was recorded in Furnace Creek (then called Greenland Ranch) in Death Valley, on July 10, 1913, some climate scientists say that reading was not verified. The following day cooled to 129 °F. Two days later, Stovepipe Wells, also in Death Valley recorded a low temperature of 107.7 °F breaking the record for the previous highest minimum ever recorded of 107 °F at Furnace Creek. No stranger to extremes, Death Valley is one of the hottest and driest places on Earth due to the shape of the valley and its location downwind of mountain ranges (Fig. 4.2).

Although most people would imagine California's temperatures to be mild with little seasonal variations, it is quite different once one ventures away from the coast. Searing heat in summer and freezing temperatures in winter occur in parts of the Mohave Desert, while snow can last well into July in the High Sierras. The notion that one can surf and ski on the same day, if one wanted, is not too far-fetched. The range between summer and winter temperatures increases as one travels further inland.

4.1   Factors that Control Temperature

In this large state, temperatures are controlled by several factors, providing a wide range of conditions throughout the state. Among the most important factors are latitude, topography, elevation, land use and distance and exposure to the Pacific Ocean.

4.1.1   Latitude

Latitude influences temperatures mainly through seasonal daylight hours and intensity of sunlight (solar altitude). Temperatures tend to cool from south to north, as the noon sun altitude is higher by almost 9½° throughout the

Fig. 4.2 Map with major cities and topography (United States Geological Survey)

year at the southernmost portion of the state compared to the northernmost. Temperatures in coastal locations change as fast as home prices from the warmer San Diego region to the cooler Crescent City area near the Oregon border. Both are moderated by marine air masses, but San Diego has an annual average temperature of 64 °F compared to 52 °F at Crescent City. Similar decreases in temperature from south to north are observed in inland regions of the state.

Latitude also dictates the length of daylight hours. Along the southern border, near 32 degrees north latitude, daylight increases from 10 h on December 21 (winter solstice) to 14 h and 18 min on June 21 (summer solstice), although these are apparent daylight times since the atmosphere bends light giving us six extra minutes before actual sunrise and after actual sunset. At Crescent City, 41.8° north latitude, winter solstice has nearly one less hour of daylight, 9 h, 17 min increasing to 15 h, 10 min at the summer solstice, nearly one more hour than extreme southern California.

4.1.2 Topography and Elevation

Topography is also a big player in varying temperatures. Large contrasts within short distances can be experienced simply by climbing to higher elevations. On a summer day, driving from the low desert up to the surrounding San Bernardino Mountains or from the sizzling Central Valley up into the Sierras, temperatures may range from over 104 to 68 °F. Since temperatures decrease by about 3.5 °F per 1000 feet of elevation gain, it is not surprising that higher locations are usually much cooler within the same air mass. However, if there are mountains separating two locations, they can serve as a barrier to cool marine air from the Pacific Ocean or hot, dry air from the desert. Orientation can also modify values, as sunny south-facing slopes may be considerably warmer than shaded north-facing ones. This allows for a longer ski season on the northern side of mountains. Mountain-valley winds also have an effect on temperature.

4.1.3 Land Use

In addition to the factors mentioned above, the human input is also an important aspect of state temperatures. While most of the state is undeveloped or rural, there are large swaths of agricultural lands as well as sprawling urban sectors. In agricultural areas, by replacing natural vegetation with crops and orchards, that are mostly irrigated with heavy use of water during the hot, rainless summers, temperatures are modified both day and night. Irrigation tends to cool the temperatures during the day, yet warm the air during the night. In cities, surfaces are mostly impervious to rainfall and absorb more sunshine than natural surfaces leading to the urban heat island effect. Additionally, tall buildings that comprise "urban canyons" effectively prevent the escape of heat at night. Cities like Los Angeles have warmed over 5 °F in the last century as more and more open space is swallowed up in houses and pavement. (More on agriculture and urbanization effects are in Chap. 13 and later in this chapter.)

4.1.4 Pacific Ocean Influence (See also El Niño, Chap. 5)

Coastal locations are directly influenced by the moderating effect of the Pacific Ocean. The cold California Current moving toward lower latitudes is further cooled by upwelling, bringing colder waters from the depths of the Pacific to the surface. Along the northern and central California coast, summer surface sea temperatures are the coldest on the western shores of the United States. Breezes on Fisherman's Wharf in San Francisco can make July feel wintery cold while across the Bay, east of the Oakland Hills, temperatures may exceed 90 °F. These large differences occur over a mere 15 miles. Similarly, early June temperatures along the southern beaches of California may be spring-like cool, with morning low clouds "burning" off by afternoon. However, inland temperatures gradually increase until they become fiery hot approaching the southeastern deserts. Areas sheltered from the cool ocean breezes are generally quite hot in the summer months. Tourism jumps during the summer along the central coast, where it may be foggy and chilly, as Central Valley or inland residents escape the sweltering heat.

During the summer, a cool breeze from the Pacific Ocean moderates temperatures, not just along the coast but sometimes well inland. The "delta breeze", allowed by a gap in the coastal mountains east of the San Francisco Bay, keep nights cool in Sacramento, which is over 80 miles inland. In winter, temperatures cool dramatically as you travel east, while coastal plains remain mild.

Since the Pacific Ocean is the largest geographic feature on the planet, it deserves more discussion. It is the main source of moisture west of the Rockies and major contributor to practically all the rain and snow that falls on the state. The Pacific Ocean is also the largest storehouse of incoming energy from the sun. The coastal sea surface temperatures greatly influence climate along the west coast. Often coastal city temperatures mirror those of the sea surface nearby. Since the ocean lags behind the continents in heating and cooling, California water temperatures slowly reach their peak in late summer and early fall, then begin cooling and cool continuously well into February and March. Similarly, coastal cities have their seasonal maxima in late August or September but minima in December or January when the oceanic influence is not as strong as it is in the summer months. For example, on average, the warmest day of the year in San Francisco is September 24. This serves as an example of the seasonal lag (Fig. 4.3). The Santa Monica climograph (Fig. 4.4) shows a similar seasonal lag of temperature.

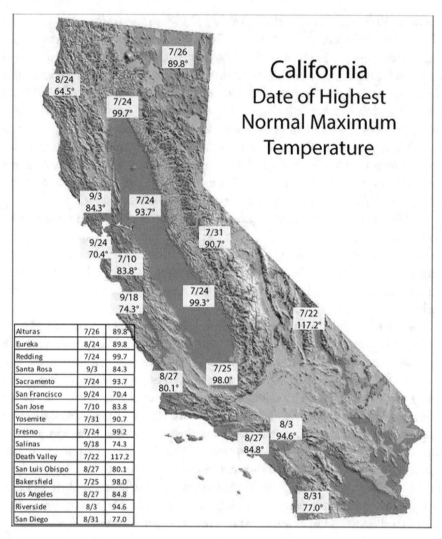

Alturas	7/26	89.8
Eureka	8/24	89.8
Redding	7/24	99.7
Santa Rosa	9/3	84.3
Sacramento	7/24	93.7
San Francisco	9/24	70.4
San Jose	7/10	83.8
Yosemite	7/31	90.7
Fresno	7/24	99.2
Salinas	9/18	74.3
Death Valley	7/22	117.2
San Luis Obispo	8/27	80.1
Bakersfield	7/25	98.0
Los Angeles	8/27	84.8
Riverside	8/3	94.6
San Diego	8/31	77.0

Fig. 4.3 Average date of the warmest normal summer temperature (Null, ggweather.com, data from NCEI, NOAA)

The mitigating influence of the Pacific Ocean gives California coastal communities some of the most temperate weather in North America, making it an attractive destination for visitors and settlers. The upwelling of cold water along the coast makes summertime water temperatures on northern and central California beaches colder than those in Washington and Oregon. San Francisco, for example, only sees a range of nine degrees between January's mean temperature of 52 °F degrees and July's mean of 61 °F. It is this summertime chill that has caught many tourists in shorts and t-shirts

Monthly Climate Normals (1991–2020) – SANTA MONICA PIER, CA

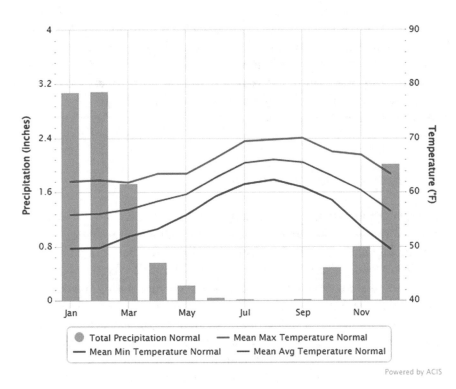

Fig. 4.4 Santa Monica climograph. Average high temperature is indicated by the red line and low temperature by the blue (Applied Climate Information System, 2022)

when they should be wearing pants and sweaters. The range between summer and winter temperatures increases the farther one travels from the coast. This means that during the summer, coastal communities are often 30–40 °F cooler than cities sometimes only a few miles farther inland.

When the North Pacific High, (an area of high atmospheric pressure), is strong, the California Current is also strong and coastal sea surface temperatures remain cool. If the high weakens, there may be a countercurrent moving from the south along the immediate coast bringing warmer water into southern California. During El Niños, even more southerly water moves north, reducing the summer ocean breeze and bringing warmer temperatures to the coast.

4.2   Seasonal Patterns

In defining the seasons, we will use the meteorological definitions. Winter is December, January, and February; Spring—March through May; Summer—June through August and fall—September through November. Much of California has a definite lag in summer temperatures compared to the rest of North America. For example, in coastal locations, the warmest month tends to be in August or September. As mentioned earlier though, winter sees its coldest temperatures in December or January much like the rest of North America.

4.2.1   Annual Temperature Range

The average annual temperature range (difference between warmest and coldest month—usually shown as the difference between the average temperatures in July and those in January) generally shows the continental effect of increases with distance from the coast (Fig. 4.5). As the amount of moisture in the air decreases and distance from moderating oceanic influences increases, annual temperature range increases. Temperature range also is reflected in latitude as higher latitudes receive greater daylight hours in summer, yet shorter hours in winter than further south. In general, differences between summer and winter temperatures are less than 18 °F along most of the state's coastline while differences increase to over 45 °F in the desert southeast as well as the extreme north and northeast interior. At similar latitudes, at distances inland, especially where mountains interrupt the oceanic influence, temperature range increases dramatically. In southern California, Santa Monica, along the coast, has a seasonal (July–January) range of 8.5 °F while San Bernardino nearly 80 miles directly east has a 25 °F range. Even further east in the dry desert of Palm Springs that range increases to 44 °F. Similar differences can be seen in northern California where San Francisco shows an 8 °F difference between the coldest and warmest month, while at Sacramento, just 80 miles to the east, this range increases to nearly 30ºF. Farther north, Redding shows a range of 36 °F. Being farther from the Pacific Ocean influence than Sacramento, Redding's summer temperatures are warmer than Sacramento's despite being father north.

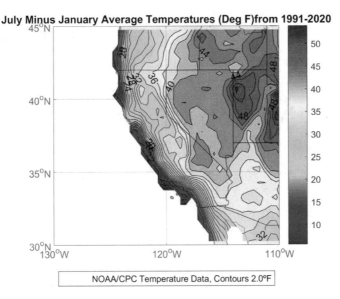

Fig. 4.5 Average temperature range, difference between July and January means (NOAA Climate Prediction Center)

4.2.2 Summer

Summer temperatures in California cover a wide range of values. From the intense heat of the southeast desert areas to the cool coast and mountains, just about any kind of summer temperature can be found.

Summer temperatures in coastal areas are relatively cool. The Pacific Ocean is a major influence on coastal areas. Ocean temperatures on the coast are quire chilly, ranging from the low 60s near Los Angeles to the low 50s near San Francisco. As a result, the city of San Francisco has an average July temperature of just 60 °F, while Los Angeles International Airport (LAX), near the coast, has an average July temperature of 69.5 °F. Things change rapidly away from coastal areas. For example, Sacramento, 80 miles east of San Francisco, has an average July temperature of 76 ºF. Riverside, a little over 60 miles east of LAX, has an average July temperature of 79 °F. (We will see less dramatic differences in winter.)

Summer temperatures are the warmest in the desert areas of southeast California. Palm Springs has a mean July temperature of 94 °F with an average high of nearly 109 °F.

Cooler temperatures can be found farther north along the coast and in mountainous regions. Eureka, on the far north coast, has an average July temperature of just under 58 °F while South Lake Tahoe at an elevation of just over 6000 feet has an average temperature of 61.5 °F.

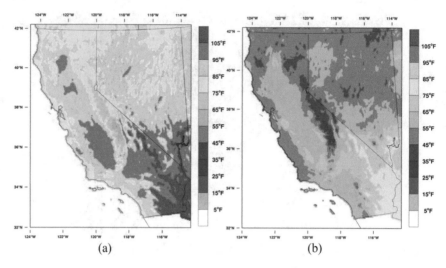

Fig. 4.6 Mean July Temperature **a** maxima and **b** minima (PRISM, Oregon State University)

During the summer, a cool breeze from the ocean moderates temperatures not just along the coast, but sometimes far inland. The "delta breeze" allowed by a gap in the coastal mountains east of San Francisco Bay keeps nights cool in Sacramento, over 80 miles inland.

Figure 4.6 shows the average July maximum and minimum temperatures across the state. July tends to be the warmest month of the year for inland regions, while regions near the coast warm up gradually, as does the Pacific, with August or September being the warmest month.

4.2.3   Fall

Fall is a transition season, when in general, temperatures begin to cool. However, many coastal locations experience their warmest weather in early fall. Fall temperatures can be quite summer-like as coastal waters reach their peak surface temperatures in early fall. Coastal cities, like Los Angeles record more heat waves in September, than August, with the highest temperatures even coming as late as October in some years. On September 27, 2010, Downtown Los Angeles hit its all-time record high of 113 °F! Normally cool, foggy Monterey Bay warmed to 103 °F as late as October 8, 1996. Only two other times, besides 1996, that Monterey hit over 100 °F on two consecutive days, both in October 1961 and 1980. Climatologically, the warmest month in San Francisco is September with an average temperature of 65.3 °F. Although the warmest month of the year at LAX is August, September is warmer than both June and July. Inland areas like Sacramento report July as their warmest month and show a distinct cooling beginning in September.

4.2.4 Winter

In winter, we find the coldest temperatures across the state at the very end of December or very early January. Winter temperatures are mild along the coast, and except for mountainous areas, we do not see the dramatic differences between inland and coastal locations that were evident in summer. The differences, we do see, however, are reversed, with warmer temperatures along the coast. In the coldest months of December and January, San Francisco is about 4 °F warmer than Sacramento. The desert areas are delightfully pleasant in winter with Palm Springs having an average high temperature of 70 °F. At just over 6000 feet in elevation, South Lake Tahoe experiences December and January temperature averages slightly below freezing at 30.5 °F making for excellent skiing in the surrounding mountains (Fig. 4.7).

4.2.5 Spring

In spring, we see warming conditions throughout the state, but there is a definite lag in warming along the Pacific Coast, due to the slow warming of the ocean. Coastal spring temperatures may be winter-like on occasion, with fog and low clouds or late season rains keeping temperatures on the cool side. For inland locations, particularly in the desert southeast, spring temperatures rise quickly with the approaching sizzling summer heat.

4.3   Diurnal (Daily) Temperature Range

Diurnal temperature ranges show a similar trend to the annual. It also is influenced by distance inland and to a lesser extent elevation. The almost always moist, or cloudy, coast retains heat at night reducing overnight cooling, while drier inland or mountainous regions cool quickly with less water vapor to absorb Earth's radiated heat. Comparing Santa Monica with San Bernardino again, the average diurnal range in July is only 7 °F on the coast, but 33 °F inland. Similarly in July, San Francisco shows close to a 7 °F difference between the high and low temperature, while the range at Sacramento is nearly 35 °F.

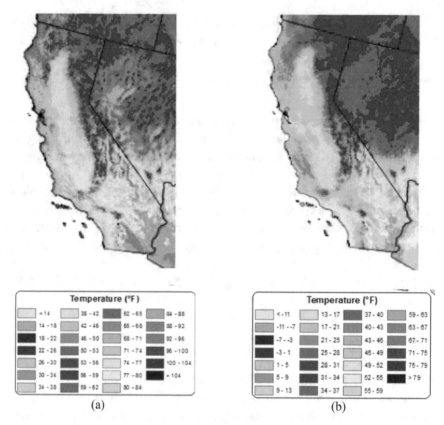

Fig. 4.7 Mean January Temperature **a** maxima and **b** minima (PRISM, Oregon State University)

4.4   Temperature Extremes

High mountain locations often hold the record cold daily temperatures for the United States. Bodie, a ghost town and state historical park in the Sierras at 8379 feet has an average of 303 nights below freezing per year. It rivals Utqiagvik, Alaska's 315 nights. In fact, *no* calendar month has ever been completely frost-free there, with the fewest nights below freezing being two in the exceptionally warm August of 1967. Bodie's record high temperature of 91 °F was set on July 21, 1988, while the record low of − 36 °F was set on February 13, 1903, which also saw the lowest maximum temperature of − 4 °F.

It is not uncommon for California to have both the nation's highest and lowest temperatures on the same day. Often during the summer, while Death Valley or a nearby location claims the hot spot of the contiguous United

States, Bodie may be the coldest, usually dipping below freezing. However, California's coldest temperature ever recorded was at Boca Reservoir at 5532 feet above sea level, just 4 miles northeast of Truckee, it was −45 °F on the morning of January 20, 1937. Boca Reservoir lies in a valley, so cold air drains off the surrounding mountains.

Following the destructive Whittier Narrows Earthquake east of downtown Los Angeles on October 3, 1987, temperatures reached a record 108 °F downtown. Downslope Santa Ana winds often elevate temperatures in early fall to 100 °F or higher (see Chap. 8).

Death Valley, in the Mohave Desert, averages 138 days per year with temperatures of at least 100 °F. It's not uncommon for the minimum to stay above the 100-degree mark. (WRCC) In winter, the same station records an average of 12 days below freezing. Although some people may not believe the unofficial 133 °F that occurred June 17, 1859 in Santa Barbara during a sudden sundowner (see Chap. 8 winds), it was the third-hottest temperature ever recorded on Earth (*It Happened in Old Santa Barbara* by Walker A. Tompkins).

4.4.1 Cold Waves

A cold wave is a rapid fall in temperature within a 24-h period. The precise criterion for a cold wave is determined by the rate at which the temperature falls, and the minimum to which it falls. This minimum temperature is dependent on the geographical region and time of year. Most cold waves are recorded when there is agricultural or property damage, such as with frost. Since the major impacts from cold waves are in agriculture, the effects of cold waves and frost are discussed in Chap. 13 (agriculture).

4.4.2 Heat Waves

The desert southeast and Central Valley are both hot in the summer months. But it can be unusually hot practically anywhere in California. These heat waves can be unpleasant to deadly for both animals and plants. In the United States, the National Weather Service suggests early warning when the daytime heat index (including adjustment for humidity) reaches 105 °F and a nighttime minimum temperature of 80 °F persists for at least 48 h. Heat index is a calculated formula that combines temperature and humidity to estimate how hot most people would feel under dry conditions (Fig. 4.8). For example, the chart shows that when the temperature is 90 °F with a humidity of 70%, it feels as if it were 105 °F under dry conditions. When feeling warm, people perspire water through pores in their skin. As this water evaporates into unsaturated air, the process consumes heat (called latent heat) taking this heat away from one's skin. As air humidity increases though, it becomes more difficult to have evaporative cooling by perspiration. People feel hotter in warm, humid conditions since evaporation is more difficult. The higher the humidity, the warmer it feels to most people.

Heat waves in California are historically dry and tolerable. The temperature warms up during the day and normally cools off at night allowing plants and animals to recuperate and get ready for another day of sizzling heat. However, this traditional type of heat wave, typical for our semi-arid Mediterranean climate, has increasingly tended to be more humid and more often warmer at night since the 1980s.

Humid heat waves begin with higher temperatures in the morning and tend to reach higher temperatures during the day, lasting longer than their dry counterparts. The trend toward more humid, more intense and longer-lasting heat waves in California was shown in the July 2006 heat wave, an event of unprecedented impact on human health in the state. A record-breaking heat wave occurred in July 2006. This heat wave was most intense from July 16 until July 26. New high-temperature records were set at seven stations. Minimum temperatures were also affected with 11 stations reporting

National Weather Service
Heat Index Chart

Temperature (°F)

Relative Humidity (%)	80	82	84	86	88	90	92	94	96	98	100	102	104	106	108	110
40	80	81	83	85	88	91	94	97	101	105	109	114	119	124	130	136
45	80	82	84	87	89	93	96	100	104	109	114	119	124	130	137	
50	81	83	85	88	91	95	99	103	108	113	118	124	131	137		
55	81	84	86	89	93	97	101	106	112	117	124	130	137			
60	82	84	88	91	95	100	105	110	116	123	129	137				
65	82	85	89	93	98	103	108	114	121	128	136					
70	83	86	90	95	100	105	112	119	126	134						
75	84	88	92	97	103	109	116	124	132							
80	84	89	94	100	106	113	121	129								
85	85	90	96	102	110	117	126	135								
90	86	91	98	105	113	122	131									
95	86	93	100	108	117	127										
100	87	95	103	112	121	132										

Likelihood of Heat Disorders with Prolonged Exposure and/or Strenuous Activity
Caution   Extreme Caution   ■ Danger   ■ Extreme Danger

Fig. 4.8 Heat index uses a formula that includes humidity and temperature to estimate how warm most people feel due to the humidity which would be equivalent to higher temperature under dry conditions (National Weather Service, NOAA)

their highest daily minimum temperature ever. Incredible heat was reported in parts of southern California. On July 22, Woodland Hills recorded the highest temperature ever for Los Angeles County at 119 °F. This record, however, was broken a few years later with the same location reporting 121 °F on September 6, 2020. Most of the deaths from hyperthermia occurred in inland counties, which were the hottest, while the highest morbidity (illness) was along the less accustomed coast. One mortality study found that each 10-degree Fahrenheit increase from day to day led to a 9% increase in daily deaths. Highest mortality was associated with the highest temperatures that occurred in Imperial and Fresno counties. The impact on farm animals and agriculture was also enormous, with the death of more than 25,000 cattle and 700,000 fowl. State-wide energy consumption on July 24, 2006, set an all-time record at over 50,000 MW. The September 2020 heat record though brutal was luckily shorter in duration and was characterized by lower humidity than the 2006 event.

The cause of humid heat waves requires high pressure in the Great Basin and low pressure off California's coast. This draws warm moist air from the south impacting California. Adding to that, coastal waters west of Baja California, an important source for humid air, have become unusually warm in recent decades.

The greatest per capita health impact from heat waves generally occurs on the north and south coasts as well as the Central Valley. Because of its large population, the south coast region experiences the highest number of hospitalizations.

The time of year heat waves occur greatly influence their impacts. Mid-season heat waves have a greater effect in the Central Valley, while early season events in coastal California have more impact because of the large deviation from the cool cloudy conditions ("May Gray" and "June Gloom".) that normally prevail. Heat waves during this part of the season come as a surprising shock contrasted to typical conditions.

Heat waves in the Los Angeles area are becoming more frequent and longer in duration since the 1900s. While heat waves (90 °F or above for three days or more) only lasted three to seven days in the early decades, they now have exceeded ten days and even two weeks. While Los Angeles hot days (90 °F or above) have increased (Fig. 4.9), cold days have decreased. On September 7, 1955, an eight-day run of 100 °F heat in Los Angeles, finally ended after causing 946 fatalities. More recently, during September 2022, Los Angeles experienced eight days running with a high temperature greater than 90 °F with the temperature reaching 100 °F on two of the days. The higher humidity made it feel even hotter.

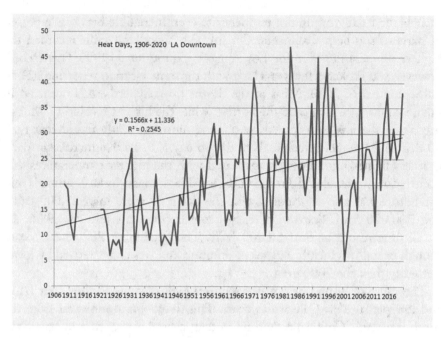

Fig. 4.9 Heat days (90 °F or above) are increasing in downtown Los Angeles (LaDochy, data from NOAA, NWS)

4.5   Urban Heat Island

Cities tend to be warmer than their surrounding rural areas due to the urban heat island (UHI) effect. This results from the built-up areas absorbing more solar radiation than rural areas and then reradiating the heat stored during the day. A nighttime satellite image of Los Angeles reveals the extensive built-up areas as well as the undeveloped land (Fig. 4.10). The effect was first described by Luke Howard in 1818 for the city of London, where nighttime temperatures increased as approaching the center of the urban core. Most cities, even towns and shopping malls, create a warmer climate than their surroundings (Fig. 4.11).

Urbanization affects weather and climate for several reasons. Typically, natural vegetation is replaced by buildings, roads and asphalt. This changes the local heat exchanges by absorbing more solar radiation than vegetation. City materials also store this heat much better than vegetation, releasing it slowly at night. Building geometry captures more incoming solar radiation and the narrow urban canyons with their tall buildings blocks more outgoing ground (infrared) radiation than natural surfaces leading to higher temperatures at night. Cities also generate tremendous amounts of heat energy that

Fig. 4.10 View of the Los Angeles metropolitan region at night shows the extensive built-up areas, while dark patches, undeveloped land, are mainly rugged hills and mountains (NASA, Science)

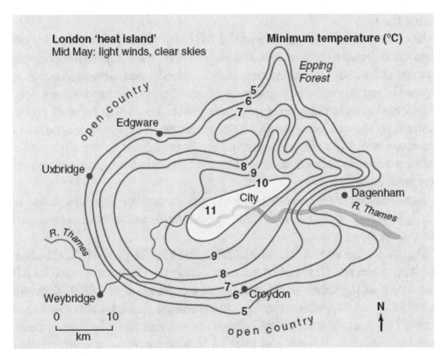

Fig. 4.11 Luke Howard's map for London's nighttime heat island (slideplayer, S. LaDochy)

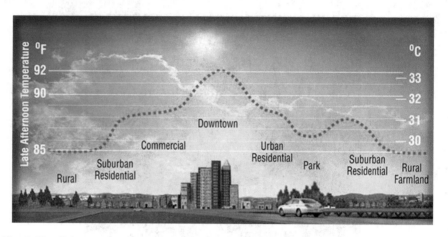

Fig. 4.12 Cities tend to be warmer than their rural surroundings, with the peak of the "island" being located in the Central Business District (Environmental Protection Agency)

contribute to the UHI. Figure 4.12 shows a typical urban temperature pattern during the day.

As cities grow, the magnitude of the UHI also grows proportionately with population. Heat islands are usually measured by the temperature difference between cities and the surrounding areas. Ground-based air temperatures can be monitored using vehicle traverses or stationary weather stations, while aircraft and satellites can measure urban surface heat using infrared sensors. Figure 4.13 shows surface temperatures as measured by a NASA overflight of Sacramento. NASA's aircraft measured the heat emitted by the city's surface. Built-up areas downtown were warmest, while parks and locations near the river were cooler.

Land use percentages for Sacramento can be seen in Fig. 4.14. Land use greatly affects temperature with built up areas being much warmer than parks and undeveloped areas.

Depending on several factors including the time of year, wind, cloudiness and land usage, the UHI can be stronger during the day or at night. The EPA estimates that the urban core for a city of a population of 1,000,000 would average 1.8–4.8 °F warmer than its surroundings. At night, this value could reach 22 °F. A study done in 1952 shows the San Francisco downtown central business district (CBD) to be at least 18 °F warmer than the green spaces of Golden Gate Park in the western part of the city (Fig. 4.15).

Cities and even towns across the state exhibit UHI effects, with the largest cities generally having the largest UHI. While inland cities such as Sacramento have a concentric ring pattern to temperatures, with one central

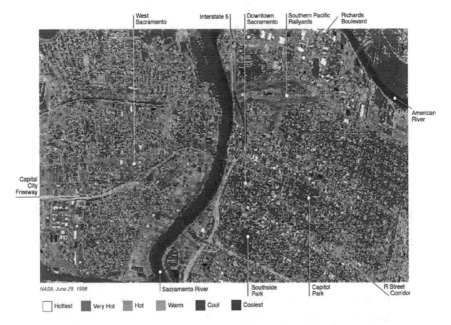

Fig. 4.13 NASA overflight of Sacramento shows how land use affects surface temperatures from very hot (red) in developed areas to cooler (blue) surfaces in area with trees or water (NASA, Science)

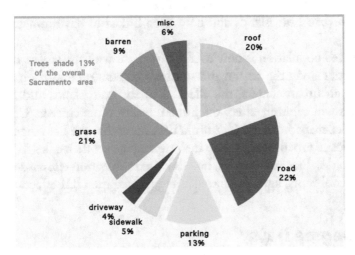

Fig. 4.14 Land use cover categories for the city of Sacramento show half of the area as residential, while trees covered only 13% (Sacramento Tree Foundation)

Fig. 4.15 San Francisco's UHI at night, with a cooler Golden Gate Park represented by the rectangle in the left of the diagram. Warmest temperatures are evident over the downtown core (Duckworth and Sandberg, Published (1954) by the American Meteorological Society)

heat island, the Los Angeles metropolitan region has been described as an archipelago of heat islands with several city cores. Temperature can also vary inside a city. Some areas are hotter than others due to the uneven distribution of heat-absorbing buildings and pavements, while other spaces remain cooler as a result of trees and greenery. These temperature differences constitute intra-urban heat islands and have been linked to environmental justice issues.

The state's population is now over 90% urban, with the largest concentrations in cities along the central and southern coasts. As population continues increasing in urban centers, the effect of urban heat islands and increasing heat waves will certainly affect the human health of its citizens. Urban Heat Island Index maps from the California EPA show that the Los Angeles region has significant urban heat islands in comparison to other cities in the state. However cities, like Los Angeles, have adopted mitigation efforts, such as tree planting, cool roofs, and cool pavements to lessen the UHI effect.

4.6   Degree Days

When hearing or seeing the daily weather report, depending on the season, you may have come across the term heating degree day (HDD) or cooling degree day (CDD). What exactly is a degree day and what is its purpose? The most common use of degree days is for tracking, or estimating, energy use as there is a linear relationship between temperatures and energy consumption. If the temperature mean for a day is below 65 °F, we subtract the mean from

Table 4.1 Heating degree days and cooling degree days for selected locations (EPA)

CA cities	Heating degree days	Cooling degree days	Latitude
Crescent city	4762	4	41.5
Sacramento	2618	1178	38.3
San Francisco	2653	163	37.4
Fresno	2346	2124	36.5
San Diego	1226	720	32.4
Palm Springs	793	4343	33.5

65 and the result is ***heating degree days***. On the other hand, if the temperature mean is above 65 °F, ***cooling degree days*** are calculated by subtracting 65 from the average daily temperature. The hotter or colder the temperatures, the more air conditioner cooling or furnace heating would be needed to make your home or office comfortable. For example, a day that records 10 HDD would require twice as much energy to heat a building as a day that recorded 5 HDD.

4.6.1 Heating Degree Days

As temperatures plunge, the heating degree days increase, as do heating costs. In Table 4.1, six California cities are compared. Of these San Diego is the mildest, though not as warm as inland locations in summer, while Crescent City, although coastal has the highest HDD units. Central Valley stations show similar values with some increases as you go north. High Sierra and other mountainous locations would show the highest HDDs in the state (Fig. 4.16).

4.6.2 Cooling Degree Days

Cooling degree days reflect mainly summer heat and air conditioning costs. With regular ocean breezes, coastal cities have their own natural air conditioner and show lower CDD units. Crescent City to the far north is cool throughout the year, while San Diego has warmer coastal waters and warmer temperatures. Inland the summer heat results in high cooling needs, with again a cooler overall CDD total to the north. Desert locations in the southeast part of the state have the highest cooling degree days of the US with totals near 10,000 CDDs (Fig. 4.17).

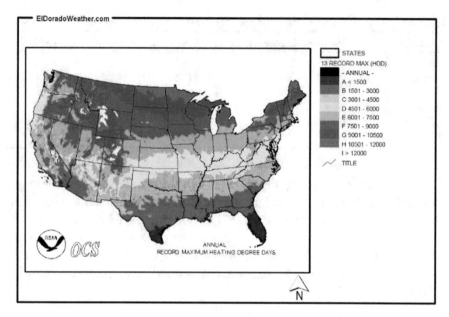

Fig. 4.16 Annual average heating degree days (Eldorado Weather, data from NOAA)

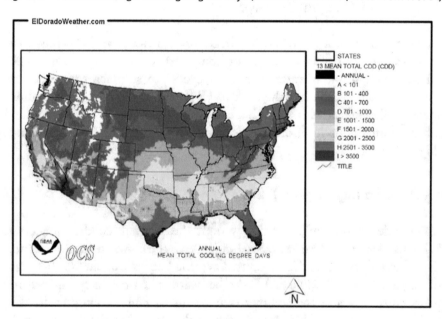

Fig. 4.17 Annual average cooling degree Eldorado Weather, data from NOAA

Bibliography

Applied Climate Information System. (2022). *Santa Monica climograph*. https://sca cis.rcc-acis.org/

Bohr, G. S. (2009). Trends in extreme daily surface temperatures in California, 1950–2005. *APCG Yearbook, 71*, 96–119.

California Department of Water Resources. (2015). *California's most significant droughts: Comparing historical and recent conditions*. Department of Water Resources, Sacramento. https://cawaterlibrary.net/document/californias-most-sig nificant-droughts-comparing-historical-and-recent-conditions/

Christy, J. R., Norris, W. B., Redmond, K. T., & Gallo, K. P. (2005). Central California: Opposing temperature trends, valley vs. mountains. In *16th conference on climate variability and change, American meteorological society*, P1.23, January 8–14, San Diego, California.

Clemesha, R. E. S., Guirguis, K., Gershunov, A., Small, I. J., & Tardy, A. (2018). California heat waves: Their spatial evolution, variation, and coastal modulation by low clouds. *Climate Dynamics, 50*(11–12), 4285–4301. https://doi.org/10. 1007/s00382-017-3875-7

Duckworth, F. S., & Sandberg, J. S. (1954). The effect of cities upon horizontal and vertical temperature gradients. *Bulletin of the American Meteorological Society, 35*(5), 198–207. https://doi.org/10.1175/1520-0477-35.5.198

Eldorado Weather. (2022). *Heating and cooling degree days*. https://www.eldoradow eather.com/

Environmental Protection Agency. (2022). *Climate change indicators: Heating and cooling degree days*. https://www.epa.gov/climate-indicators/climate-change-indica tors-heating-and-cooling-degree-days

Environmental Protection Agency. (2017). *Learn about heat islands*. https://19janu ary2017snapshot.epa.gov/heat-islands/learn-about-heat-islands_.html

Gershunov, A., Cayan, D. R., & Iacobellis, S. F. (2009). The great 2006 heat wave over California and Nevada: Signal of an increasing trend. *Journal of Climate, 22*(23), 6181–6203. https://doi.org/10.1175/2009jcli2465.1

Guirguis, K., Gershunov, A., Tardy, A., & Basu, R. (2014). The impact of recent heat waves on human health in California. *Journal of Applied Meteorology and Climatology, 53*(1), 3–19. https://doi.org/10.1175/jamc-d-13-0130.1

Jet Propulsion Laboratory, California Institute of Technology. (2018). *ECOSTRESS maps LA's hot spotts*. https://ecostress.jpl.nasa.gov/news/1nasas-la-hot-spots-image

Knowlton, K., Rotkin-Ellman, M., King, G., Margolis, H. G., Smith, D., Solomon, G., Trent, R., & English, P. (2009). The 2006 California heat wave: Impacts on hospitalizations and emergency department visits. *Environmental Health Perspectives, 117*, 61–67.

LaDochy, S. (2022). *Heat days in Los Angeles*. Data from NOAA, NWS. https:// www.weather.gov/wrh/climate?wfo=lox

LaDochy, S. (2013). California is heating up, but it's oh so much worse in the cities. In *Presentation at Conference on Global Modernities.* https://slideplayer.com/slide/7493963/

LaDochy, S., Medina, R., & Patzert, W. (2007). Recent California climate variability: Spatial and temporal patterns in temperature trends. *Climate Research, 33,* 159–169.

NASA Science. (1998). *Cities getting ready for the next heat wave.* https://science.nasa.gov/science-news/science-at-nasa/1998/essd20nov98_1

National Weather Service, NOAA. (2022). *What is the heat index?* https://www.weather.gov/ama/heatindex

Nemani, R. R., White, M. A., Cayan, D. R., Jones, G. V., Running, S. W., Coughlan, J. C., & Peterson, D. L. (2001). Asymmetric warming over coastal California and its impact on the premium wine industry. *Climate Research, 19,* 25–34.

NOAA Climate Prediction Center. (2023). https://www.cpc.ncep.noaa.gov/

Null, J. (2022). *Date of highest maximum temperatures.* https://ggweather.com/links.html

Null, J., & Mogil, M. (2010). The weather & climate of California. *Weatherwise, 63*(2), 16–23.

Office of Governor Gavin Newsom. (2021). *Governor Newsom signs Climate Action Bills, outlines historic $15 billion package to tackle the climate Crisis and protect vulnerable Communities.* https://www.gov.ca.gov/2021/09/23/governor-newsom-signs-climate-action-bills-outlines-historic-15-billion-package-to-tackle-the-climate-crisis-and-protect-vulnerable-communities/

Patzert, W. C., LaDochy, S., Ramirez, P., & Willis, J. K. (2016). Los Angeles weather station's relocation impacts climatic and weather records. *The California Geographer, 55,* 41–52. http://scholarworks.csun.edu/handle/10211.3/170953

Parzybok, T. W. (2005). *Weather extremes in the West.* Mountain Press.

PRISM Climate Group, Oregon State University. (2022, December 2). *30-year normals.* https://prism.oregonstate.edu/normals/

Sacramento Tree Foundation. (2022). *Sacramento land use cover.* https://sactree.org/

Southwest Climate Science Center. (2015). *California heat Waves: A new type of heat wave.* https://www.swcasc.arizona.edu/sites/default/files/HeatWaves.pdf

Tamrazian, A., LaDochy, S., Willis, J., & Patzert, W. (2008). Heat waves in southern California: Are they becoming more frequent and longer lasting? *Yearbook of the Association of Pacific Coast Geographers, 70,* 59–69.

Tompkins, W. A. (1976). *It happened in Old Santa Barbara.* Santa Barbara National Bank.

United States Geological Survey. (2022). *The National map.* https://www.usgs.gov/programs/national-geospatial-program/national-map

Western Regional Climate Center. (2018). *California climate tracker.* https://wrcc.dri.edu/monitor/cal-mon/

5

The Atmosphere–Ocean Influence

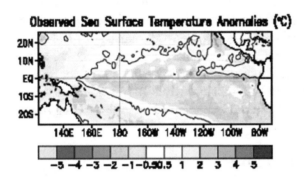

Fig. 5.1 Sea surface temperature anomalies on 14 September 2022 showing La Niña conditions (Climate Prediction Center)

There are more things in heaven and earth, Horatio, than are dreamt of in your philosophy.—William Shakespeare (Hamlet)

Severe drought in India, four tropical cyclones in Tahiti, hurricanes in Hawaii—these events marked the very strong El Niño of 1982–83. In California, this El Niño was accompanied by record rainfall. Then, early in the summer of 1997, it became apparent that another intense El Niño was on the way, and concern rose from Australia to India to California. As India braced

© The Author(s), under exclusive license to Springer Nature
Switzerland AG 2023
S. LaDochy and M. Witiw, *Fire and Rain*,
https://doi.org/10.1007/978-3-031-32273-0_5

for drought, Tahiti prepared for more tropical cyclones. Although aspects of the 1997–98 El Niño were similar to those of 1982–83, the Indian monsoon rains arrived on time, and Tahiti experienced no tropical cyclones. However, southern California did have another record year for rainfall (Fig. 5.1).

5.1 Atmosphere–Ocean Interactions: What Is Known and Not Known?

The words of Hamlet can be applied to our knowledge of the interaction between the atmosphere and the ocean. We know of many ways they influence each other, but we still don't know all the effects they have on our weather and climate. These interactions occur over different areas and involve the exchange of heat energy and moisture between the ocean and the atmosphere. Where atmospheric winds drive ocean currents, there is also an exchange of kinetic energy (energy that comes from motion). In turn, these interactions affect atmospheric circulations as well as precipitation and temperature patterns. One such example that has major impacts on California is the Pacific Decadal Oscillation (PDO), but the most well-known of these is the El Niño Southern Oscillation (frequently referred to as the ENSO).

5.2 What Is an El Niño?

Before looking in detail at the impacts specific El Niños have had on California, let's look at what an El Niño is, and the history behind the name.

The waters off the west coast of South America are normally very cold for their latitudes. Near Lima, Peru (12 degrees south latitude), normal ocean temperatures vary from about 61 °F in the winter to about 68 °F in the summer. The cold current of water moving from the south to the north extends nearly to the equator before turning to the west. Typically, for a period of just a few weeks around Christmas each year, this cold water is replaced by a warm current. Because of the timing, Peruvian fishermen called this local event El Niño, in honor of the Christ child. In the twentieth century, however, it was discovered that anomalously warm water temperatures in parts of the Pacific Ocean sometimes last for a year or more. Over time, the term El Niño has been adopted to describe these longer periods of ocean warming.

The diagrams (Fig. 5.2) illustrate normal conditions contrasted with El Niño conditions. The surface water temperatures off the coasts of the

Americas are shown, with coldest water depicted in blue and warmest in red. The reddish-brown depicts coastlines, with the Americas on the right side and Australia in the lower left quadrant. The diagram also shows the vertical profile of the thermocline (a transition layer between deep cold water and warmer surface water). The thermocline is shallow off the coast of South America and deep in the western Pacific during normal conditions and/or La Niñas. During normal years, the trade winds (indicated by the white arrows) blow from east to west and push warmer surface water to the western Pacific. This allows cool water off the coast of South America to upwell from the depths below to the surface.

During El Niño years, however, the trade winds weaken, and warm surface water remains near the South American coast. In very strong El Niños, the trade winds may disappear and be replaced by winds from the west as

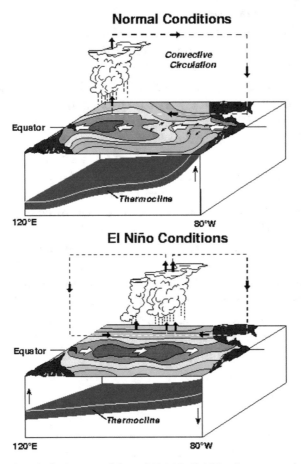

Fig. 5.2 Normal and El Niño conditions (NOAA's El Niño theme page)

shown in Fig. 5.2. Subsequently, there is a shift in weather patterns across the Pacific. This usually results in drought in parts of the western Pacific region (Australia, Southeast Asia) and above normal precipitation in the central and eastern Pacific region, especially in southern California. El Niño's impact, however, is not limited to the Pacific and its bordering land masses, but here, we will look specifically at the effects of El Niños on California.

Since 1950, 25 El Niños have been identified. Three of these—the occurrences of 1982–83, 1997–98 and 2015–16—were characterized as being very strong. In a very strong El Niño, ocean temperatures across a wide part of the central and eastern Pacific are 2.0 °C (3.6 °F) or more above the long-term average for a period of several months. For short periods, temperatures may exceed this by over four degrees Celsius above average (nearly eight degrees Fahrenheit). Another five events were identified as strong. In a strong El Niño, the criterion is an extended period of temperatures 1.5 to 1.9 °C (2.7 °F to 3.4 °F) above average.

In the El Niños of 1982–83 and 1997–98, southern California received record rainfall. Los Angeles International Airport (LAX) reported seasonal totals of nearly 26 inches and over 31 inches of rain respectively for these two seasons. (Normal seasonal rainfall for LAX is just over 12 inches.) When it became apparent that another very strong El Niño was imminent in 2015, there was widespread hope that the drought that had begun in 2011 would soon end. Although the very strong El Niño did occur, the results in southern California were disappointing. LAX received just a bit more than nine inches of rain for the 2015–16 season, showing that El Niño, although very important, is only one factor in explaining precipitation variability in California. The intensity of the 2015–16 El Niño can be seen in the Fig. 5.3. The dark red indicates positive temperature anomalies of 9 °F.

The El Niño relationship with California precipitation is stronger to the south and weakens in the northern part of the state. In some El Niño years, northern California, (along with Oregon and Washington) has below normal rainfall. When all El Niño years are considered, both San Diego and LAX accumulated an average of more than 120% of normal rainfall, while San Francisco and Eureka received very close to normal. Very strong El Niño years typically (but not always) bring above average rainfall throughout the state.

5.2.1 La Niña

When ocean temperatures in the central and eastern tropical Pacific shift to cooler than average, California precipitation usually drops to lower-than-normal amounts. This cooling of the eastern Pacific stabilizes the atmosphere

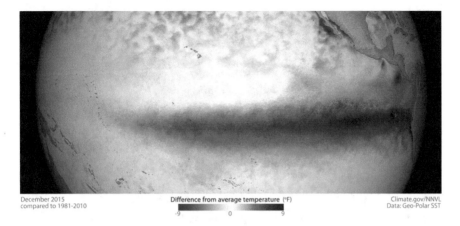

December 2015
compared to 1981-2010

Difference from average temperature (°F)

-9   0   9

Climate.gov/NNVL
Data: Geo-Polar SST

Fig. 5.3 Water temperature anomalies, as much as 9 °F above normal, in December 2015 NOAA, (Climate.gov, 2016)

and lessens storm frequencies. The term La Niña (the little girl) was adopted to contrast with El Niño (the boy child). As with El Niño years, the La Niña influence is strongest in the southern part of the state. For all La Niña years, from San Diego to San Francisco, precipitation can drop as much as 32% below what is normally expected, while in the far northern part of the state, near average conditions may prevail. Often, La Niña events follow El Niño events and vice versa. La Nada (the nothing) is now used to describe neutral water temperature anomalies (Fig. 5.4).

5.2.2 El Niño—Southern Oscillation (ENSO)

The Southern Oscillation was discovered at the same time scientists in South America began to document the local effects of El Niño in the early 1900s. A British scientist, Sir Gilbert Walker, was in India attempting to develop methods for predicting monsoon intensity. Examining world weather data, Walker found an intriguing correlation between the surface pressures on the eastern and western sides of the tropical Pacific Ocean. When pressure rises on one side of the tropical Pacific, it usually falls on the other. Walker introduced the term Southern Oscillation to refer to this east–west seesaw in the pressures of the tropical Pacific. The use of the term "southern" stems from the fact that this effect is strongest in the tropical Pacific south of the equator. The Southern Oscillation Index (SOI) is calculated using pressure readings from Tahiti and Darwin, Australia.

In the late 1960s, Jacob Bjerknes and others realized that the changes in the ocean and the atmosphere were connected, and the hybrid term El

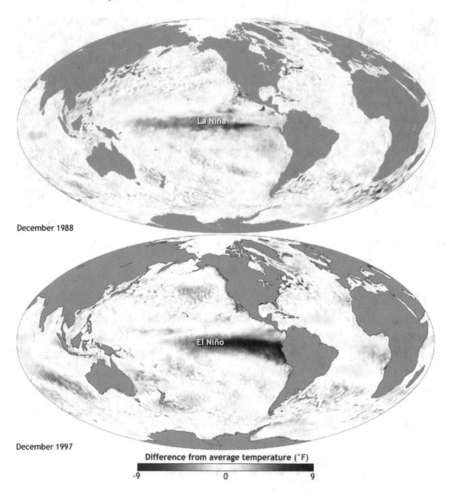

Fig. 5.4 Eastern Pacific water temperature anomalies, as much as 9 °F below normal during a La Niña (top) and as much as 9 °F above normal during an El Niño (bottom) NOAA, (Climate.gov, 2016 El Niño and La Niña: Frequently asked questions:)

Niño-Southern Oscillation (ENSO) was born. During average (or normal) conditions, with high pressure in the eastern tropical Pacific and low pressure over the equatorial western Pacific, surface winds (called trade winds) blow from east to west. Warmer ocean water piles up on the western side of the tropical Pacific, while cooler water upwells along the South American coast. Warmer water fuels wet conditions to the west and cool water reduces updrafts in the atmosphere to the east. When the trade winds are weaker than normal, the warm water usually found in the western Pacific flows eastward and El Niño events occur. This is the warm phase of the El Niño-Southern Oscillation. When conditions shift to stronger than normal trade winds, the

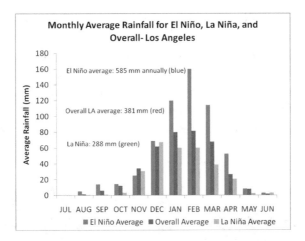

Fig. 5.5 ENSO relationship to precipitation is strongest to the south. The monthly average rainfall for Los Angeles shows higher amounts during El Niño events as well as shifting the rainfall season to later than normal (Killam et al., 2014)

circulation leads to La Niña events (cool phase of ENSO). The warm and cool ENSO phases may each last over a year (Fig. 5.5).

5.3   The Pacific Decadal Oscillation (PDO)

A longer-lasting oscillation called the Pacific Decal Oscillation or PDO that impacts both ENSO and California weather was discovered in the 1990s. Fisheries scientist, Steven Hare, coined the term "Pacific Decadal Oscillation" (PDO) in 1996 while researching connections between Alaska salmon catch totals and the North Pacific environment.

The PDO resembles the El Niño-like sea surface temperature patterns but lasts much longer, decades. There are several characteristics that differentiate the PDO from the El Niño/Southern Oscillation (ENSO). The amplitude of ENSO can be much greater than that of the PDO. In an ENSO event, sea surface temperatures can vary as much as 9 °F from long-term averages, while variations in the PDO are no more than about 1.5 °F. Twentieth-century PDO "events" persisted for 20-to-30 years, while typical ENSO events persisted for six to 18 months. Also, the climatic patterns of the PDO are most evident in the North Pacific/North American sector, while less evident in the tropics—the opposite is true for ENSO. Several independent studies find evidence for just two full PDO cycles in the past century: "cool" PDO phase occurred from 1890 to 1924 and from 1947 to 1976, while the "warm" PDO phase took place from 1925 to 1946 and 1977–1997.

Figure 5.6 shows the sea surface temperature (SST) patterns for the warm and cool phases of the PDO, while Fig. 5.7 shows the PDO monthly values from 1850 to 2020. El Niños are more numerous and amplified during warm phases of the PDO, while La Niñas are more numerous and amplified during cool phases of the PDO. Two of the strongest El Niño events of the twentieth century took place during the warm phase from 1977 to 1998, 1982–83, and 1997–98 events. More La Niña years and California droughts have occurred since 1998.

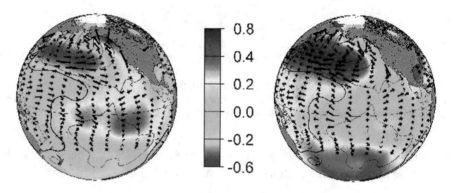

Fig. 5.6 Warm phase of the PDO is shown on the left, cool phase on the right. Note the temperature differences on the west coast of North America (Joint Institute for the study of the Atmosphere and Ocean)

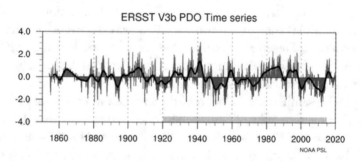

Fig. 5.7 PDO values from 1850 to 2020. Blue indicates a cool phase and red a warm phase (NOAA: Pacific Decadal Oscillation)

5.4 The North Pacific Oscillation (NPO) and Other Interactions

Since the discovery of the PDO and its well documented impacts, we have heard relatively little about the NPO. Like the Southern Oscillation, the NPO was first described by Sir Gilbert Walker in the early twentieth century. The North Pacific Oscillation (NPO) is a north–south fluctuation in winter sea level pressure over the North Pacific on monthly time scales. In 1924 Sir Gilbert Walker found that there was an oscillation between the Aleutian Low near Alaska and the Hawaiian High. During the positive phase of the NPO, there is a steeper than normal pressure gradient between a strong Aleutian Low and high pressure over Hawaii. NPO variability is associated with winter weather anomalies in North America. When the Aleutian Low is stronger than its average intensity (positive NPO phase), the North American continent tends to be warmer than the long-term normal, with less precipitation in the Pacific Northwest. There appears to be little effect on rainfall in California.

The Pacific North American (PNA) pattern is another fluctuation that is greatly influenced by ENSO and may add some to warming in California during its positive phase but with little effect on precipitation.

Several more ocean–atmosphere interactions have been discovered but appear to have negligible impacts on the weather and climate of California. These include the North Atlantic Oscillation and the Arctic Oscillation. Both are similar in structure to the NPO. However, one that has been studied recently is the Madden–Julian Oscillation (MJO) with periods of 30–60 days. It starts in the western Pacific and initially is characterized by strong thunderstorms. It loses this characteristic as it moves eastward in the tropical Pacific. Nevertheless, the disturbance that initially accompanied the thunderstorm activity may have some impact on California's precipitation.

5.5 Conclusion

Although several atmosphere–ocean interactions that impact California have been discovered, much is still not known. Long-term forecasts that incorporate the effects of these interactions can often be made with some accuracy but sometimes are off by large margins. ENSO is most often included in long-range weather forecasts for North America.

Bibliography

American Meteorological Society. (2022, December 1). *Glossary of meteorology.* https://glossary.ametsoc.org/wiki/Welcome

Bui, H. X. (2020). Madden-julian oscillation increased impacts on US West Coast, *Nature, 10,* 603–604.

Joint Institute for the Study of the Atmosphere and Ocean (2022, December 1). *The Pacific Decadal Oscillation (PDO).* http://research.jisao.washington.edu/pdo/

Killam, D., Bui, A., LaDochy, S., Ramirez, P., Patzert, W., & Willis, J. (2014). California getting wetter to the north, drier to the south: Natural variability or climate change? *Climate, 2*(3), 168–180. https://www.mdpi.com/2225-1154/2/3/168/htm

L'Heureux, M. (2014). *What is the El Niño–Southern oscillation (ENSO) in a nutshell?* https://www.climate.gov/news-features/blogs/enso/what-el-ni%C3%B1o%E2%80%93southern-oscillation-enso-nutshell

L'Heureux, M. (2019). *The Pacific-North American pattern: The stomach sleeper of the atmosphere.* https://www.climate.gov/news-features/blogs/enso/pacific-north-american-pattern-stomach-sleeper-atmosphere

Linkin, M. E., & Nigam, S. (2008). The North Pacific oscillation–West Pacific teleconnection pattern: Mature-phase structure and winter impacts. *Journal of Climate, 21,* 1979–1997.

Mantua, N. J., Hare, S. R., Zhang, Y., Wallace, J. M., & Francis, R. C. (1997). A Pacific interdecadal climate oscillation with impacts on salmon production. *Bulletin of the American Meteorological Society, 78,* 1069–1079.

NOAA. (2016). *El Niño and La Niña: Frequently asked questions.* https://www.climate.gov/news-features/understanding-climate/el-ni%C3%B1o-and-la-ni%C3%B1a-frequently-asked-questions

NOAA Physical Sciences Laboratory. (2022, December 1). *Pacific decadal oscillation.* https://psl.noaa.gov/pdo/

Null, J. (2018). *California El Nino and La Nina precipitation.* https://www.ggweather.com/ca_enso/ca_elnino.html

Null, J. (2022). *El Nino/La Nina resources.* https://ggweather.com/enso.htm

Walker, G. T. (1924). Correlation in seasonal variations in weather—A further study of world weather. *Memoirs of the India Meteorological Department, 24,* 275–332.

6

Precipitation: Too Much or Too Little

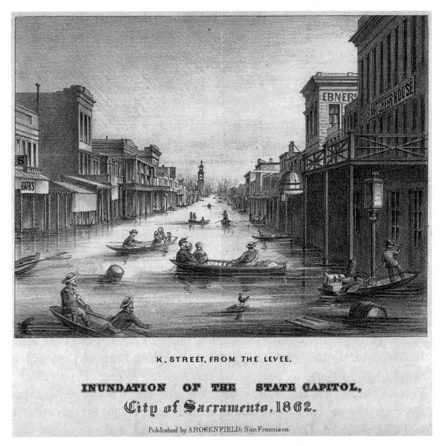

Fig. 6.1 Great 1861–62 Flood showing inundated Sacramento from *Flooding in California* (United States Geological Survey 2011)

© The Author(s), under exclusive license to Springer Nature
Switzerland AG 2023
S. LaDochy and M. Witiw, *Fire and Rain*,
https://doi.org/10.1007/978-3-031-32273-0_6

It never rains in California…but girl, don't they warn ya…it pours, man, it pours.
—Albert Hammond

Floods along rivers in California are not rare. An example is the Russian River which floods on average every other year. With about 60 inches of rain in the area annually, the steep canyons funnel water on its way to the Pacific. The town of Guerneville, in Sonoma County, north of San Francisco, has been destroyed by flooding 38 times since 1940. In 2019, a foot of rain fell on Guerneville, and 20″ fell nearby, raising the river to 13 feet over flood stage. Instead of building another dam in the popular recreation area, the town has agreed to raise buildings 10 feet. Figure 6.1 shows the Great Flood of 1861–62.

6.1   Overview

Although California is known for its sunny skies, it does rain and snow. As we saw in the introductory paragraph, the rain (and sometimes snow) can be excessive at times. Hail, drizzle, and sleet also fall. Precipitation amounts vary greatly with over 150 inches in the northwestern temperate rainforest mountains to less than two inches in Death Valley. With its length, its diverse mix of mountains, deserts, valleys, and plains, the state has the whole range of precipitation within its many climatic zones. The amount is crucial to the state with the largest population, largest agricultural produce, and the largest economy in the United States and the fifth largest globally. Normally there is plenty of water in the state for all its uses. But the state has seen devastating, long periods of droughts and even megadroughts as well as heart-breaking massive floods. The state has experienced some of the wettest weather of anywhere in the United States, including the hurricane-prone Gulf states, while long-lasting, state-wide droughts have dried up fields and led to huge wildland fires.

Rainfall (and snow) across the state varies as much as the complex topography. Fueled by the Pacific Ocean, evaporated moisture follows the westerlies throughout the state. The highest annual rainfalls are in the northern Coast Range, where annual amounts may amount to 150 inches at higher elevations. Pacific storm moisture is lifted by the Coast Range, Sierra Nevada, and other north–south mountains chains. (This is called orographic uplift where rising air is cooled and the moisture in it condenses.) On the lee of these mountains and more distant from the Pacific, there is much less precipitation.

Figures 6.2 and 6.3 shows the vast differences between orographic uplift rain and snow on the windward side of mountains and the rain shadow effects on the lee sides. In contrast to the temperate rainforest of the northwest, Death Valley and the desert southeast include some of the driest locations in North America. Death Valley, averaging 2.3 inches a year, the lowest annual total in the nation. Bagdad, also in the Mohave Desert of California, holds the United States record for the most consecutive days with no rain, 767 days running from 1912 to 1914.

Parts of California are surprisingly wet. Annual rainfall exceeds 50 inches in parts of the coast range, the west slopes of the Sierra and much of the Cascades. It drops off to about 20 inches per year in most of the lowlands surrounding San Francisco Bay and Monterey Bay. East of the coastal mountains, precipitation drops off considerably and ranges from less than eight inches per year in the southern San Joaquin Valley to more than 20 inches just

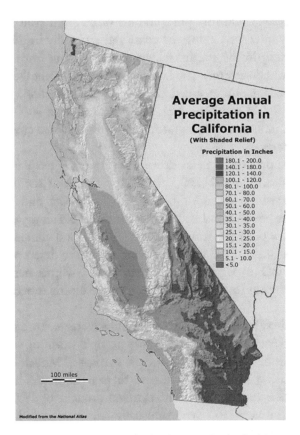

Fig. 6.2 Annual average precipitation (United States Geological Survey: The National Map)

to the north of Sacramento increasing to over 30 inches per year in the most northern part of the Central Valley. In the extreme northeast interior portion of the state, precipitation averages less than 20 inches per year (Fig. 6.2).

Nearly three-quarters of California's rain falls in the northern third of the state, while the majority of the thirsty population lives in the southern end. The demand for municipal water is greatest in the southern coastal cities, especially during the hot summers, when very little precipitation occurs.

Snow is very important to the California economy, not just for providing recreational opportunities, but for the much-needed water for agriculture and the big cities. Although very rare in the coastal cities, in parts of the Sierra, average snowfall exceeds 400 inches per year.

6.2   Seasonality of Precipitation

The seasonality of California precipitation varies from true Mediterranean (summer dry, winter wet) along the coast and especially in the southwest to a bimodal pattern in the southeast, with rains coming from winter storms and summer thunderstorms. The latter is often due to the North American Monsoon. (It is sometimes called the Southwest Monsoon, but since that name is also used in southern Asia, the preferred term is North American Monsoon.) Late summer, early fall tropical storms may also cross the southern border of the state.

Figure 6.2 shows the general decrease of annual precipitation from the northwest toward the southeast. There is also a shift in the peak of precipitation with northern locations peaking in December and January while southern locales peak around February. This shift goes along with the associated swing of the polar jet stream southward in fall and winter and retreating north in spring. Inland, Brawley and Needles, in the desert southeast, have less than five inches per year of rainfall with a winter maximum and a secondary maximum in late summer, during the North American Monsoon (Fig. 6.4).

Of course, these graphs and figures are only averages representing decades of individual events and years. Sometimes a single storm or wet period may change a year from extreme dry to average, such as a "Miracle March" saving an otherwise below average rainfall year to near normal. This occurred in 1991, after a five-year drought. The extreme drought of 1976–77 came to an abrupt end in May 1977 with unusually late showers followed by a wet 1977–78 rain season.

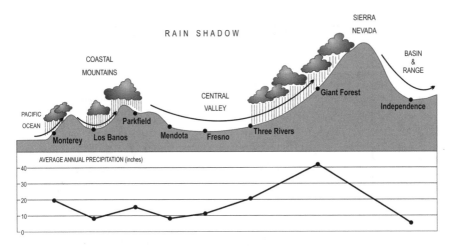

Fig. 6.3 Cross-section of central California from Monterey on the coast to Independence in the lee of the Sierra. Rainfall drops off to the lee of the coastal range and rises along the windward slopes of the Sierra. (Annual precipitation amounts in inches are from Western Regional Climate Center, NOAA. Graphics by Weldon Hiebert, adapted from Anderson (1975)

6.2.1 North American Monsoon

A true monsoon, like that in southern Asia, is defined as a seasonal reversal of winds. The North American Monsoon does not exhibit a true seasonal reversal of winds, because during the rest of the year, winds vary quite a bit. The North American monsoon is confined to the period generally between July and the end of September. This monsoon affects Mexico and Central America as well as the southwest United States. Statistically, August 8 is the peak of the season when the most rain and thunderstorms occur mainly across the southern portions of Arizona and New Mexico, sometimes spilling into southeast California. Warm, moist air moves into the southwestern United States from the Gulf of California and the Gulf of Mexico. This inflow of warm, moist air moving at low levels over the hot (unstable) desert surface air may result in numerous thunderstorms especially as air is forced upward over mountains and in complex terrain convergence zones. These sudden rains provide life-giving sustenance to natural vegetation and crops and filling reservoirs for the growing population of the American Southwest. But these thunderstorms can also lead to damaging flash floods. Other hazards include lightning strikes, damaging hail, tornadoes, and downburst windstorms. Three to five moisture surges per month during the monsoon

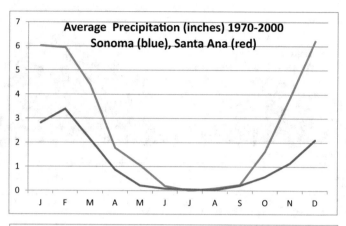

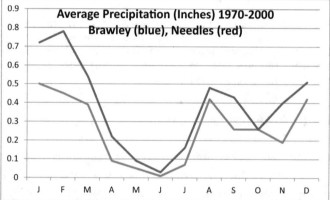

Fig. 6.4 Average precipitation. ABOVE: Sonoma (blue) annual average 31.43″, Santa Ana (red) annual average 13.63″. BELOW: Brawley (red) annual average 3.11″, Needles (blue) annual average 4.62″ (data from NOAA NCEI)

are typical. They are associated with tropical easterly waves (tropical disturbances in the easterly flow) or the occasional eastern Pacific tropical cyclone passing well to the south.

The North American Monsoon varies from year to year. If there is a trough of cool air along the eastern Pacific, the North American monsoon will be weak. In 2020, the monsoon was nearly absent, leading to extreme drought and a record wildfire season. If coastal winds are light and especially when tropical disturbances travel up the Gulf of California, more moisture arrives in the deserts of Arizona and parts of southern California.

The North American Monsoon generally occurs after water temperatures in the northern Gulf of California exceed 83.3 °F. The wettest years are associated with significantly higher temperatures (greater than 84.2 °F) while the

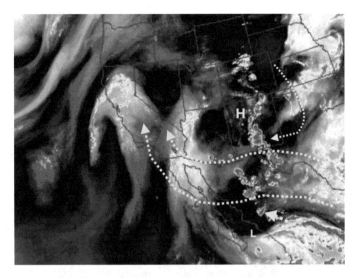

Fig. 6.5 This July 21, 2019, satellite water vapor image shows lower atmospheric flow bringing moisture into southeastern California and Arizona (NOAA National Environmental Satellite Data and Information Service)

driest years are associated with lower temperatures. The sea surface temperature (SST) in the Pacific affects when the North American Monsoons begin, with warmer temperatures resulting in a later onset (Fig. 6.5).

6.2.2 Atmospheric Rivers (ARs)

Atmospheric rivers (ARs) are long plumes of moisture bringing large amounts of water vapor from the warm subtropical Pacific to the West Coast of the United States. ARs provide much more moisture and widespread precipitation than the North American Monsoon (Fig. 6.6).

An atmospheric river that originates near Hawaii is called a "Pineapple Express" since it brings tropical moisture from near Hawaii. Atmospheric rivers frequently reach the West Coast during the winter rainy season. When they do, they drop copious amounts of precipitation. In California, ARs are responsible for up to 50% of the state's annual precipitation and 80% of major floods. AR landfalls occur in both the northern and southern portions of the state.

Atmospheric rivers in early January 2017 resulted in over 100 reports of flooding, flash flooding, or landslides in California and Nevada. Many rivers and roads across central and northern California flooded. A mudslide closed a portion of Interstate 80 near Truckee, California. High winds and

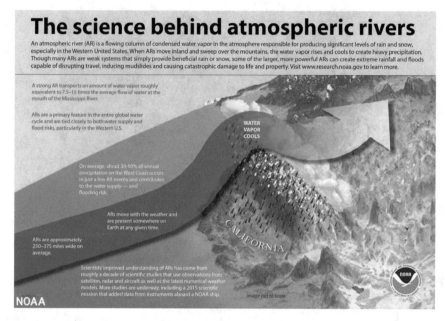

Fig. 6.6 Atmospheric rivers provide abundant moisture to the state, which is beneficial during droughts. However, too much localized heavy rains can lead to flooding and mudslides (NOAA)

avalanches occurred in the Sierra Nevada, downing trees in several northern areas. A wind gust in the Sierra Nevada Alpine Meadow reached 174 mph. One unfortunate casualty of strong winds was the loss of the famous Sequoia "tunnel tree" in California's Calaveras Big Trees State Park, estimated to be more than 1000 years old.

The state averages seven strong atmospheric river events in a water year. A larger number of ARs often result in wetter years, while fewer in drier years. Recently, atmospheric rivers were categorized by the amount of damage they produced, similar to the tornado EF-Scale. On a scale of 1 to 5 (beneficial to hazardous), the October 19–26, 2021 (Fig. 6.7), AR was deemed category 5, with record-setting precipitation throughout northern California. Sacramento set a 24-h rainfall record of 5.44 inches, nearly 70% of their water year (October 1, 2021–September 30, 2022) total. Despite the damage, the heavy rainfall and snow were welcome. However, it did not ease the severe drought of the previous two years.

Changes, however, occurred beginning in December of 2022. A series of atmospheric rivers began to impact California. The first of these only affected northern California. From December 26 through December 31, measurable rain was recorded on every day in downtown San Francisco culminating with

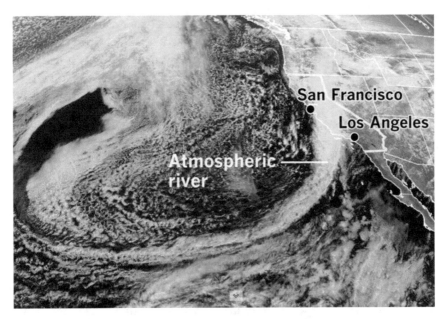

Fig. 6.7 Atmospheric river approaching California on October 22, 2021, with record-breaking rainfall in northern California (Friends of the River, image from NOAA)

a rainfall of 5.46 inches on December 31—making it the second rainiest day in over 170 years of record-keeping. The trend continued in January 2023 with over eight inches falling on the first 16 days—nearly four times the long-term average for that period. A daily record of 1.22 inches was recorded on January 10. By January 16, in just over three and on-half months downtown San Francisco had reported over 95% of its normal rainfall for the entire water year (October 2022–September 2023). Similar high rainfall amounts were seen throughout northern and central California.

In southern California, the rain was not quite as heavy, but by January 16, 2023, downtown Los Angeles had recorded three days with daily record rainfall (November 8, December 31, and January 9). Although December's rainfall was just slightly above average, the January amounts, similar to those in San Francisco exceeded four times the long-term average for the first 16 days of the month. By January 16, downtown Los Angeles had recorded 91% of its normal water year rainfall. A little farther north, Santa Barbara reported nearly eight inches of rain by January 12 with a record daily rainfall of 4.22 inches on January 9.

In the coastal mountains and hills, much higher amounts of rain occurred in places resulting in flooding and mudslides in many areas. While much of the state was under extreme (D3) to exceptional (D4) drought prior to the heavy rains, most of the state remained under moderate (D2) to severe (D3) drought. Although the rain was welcome and greatly helped the drought situation, because of the short period in which it occurred, much was lost to runoff.

During La Niña years, total precipitation amounts are usually below normal in southern and central California and slightly above normal in the northern regions. However, this was not the case during the 2022–2023 La Niña.

6.3   Precipitation Variability

Precipitation can vary greatly from year to year. In fact, California has the highest annual precipitation variability within the 48 contiguous states. The highest variability occurs in the southeast desert region where annual precipitation is typically the lowest. When averages are so low, just a few rainfall events could result in more than double the average rainfall. December 2010 was exceptionally rainy in southern California when more than 20 inches of rain fell on six days when a Pineapple Express fire-hosed southern California, mainly in San Diego County following another dry year.

Only four wet years have occurred so far in the twenty-first century for the state, the last being 2018–19. The most recent period included the historic five years drought (2012–2016) with the first three years being an exceptionally dry in the state precipitation record. Water year 2014–15 tied for the driest year on record, while 2015–16, a warm El Nino year experienced unprecedented lack of snow. Then 2016–17 brought the wettest water year on record for much of California, with widespread flooding and unprecedented atmospheric river (AR) activity. Typical of California's boom and bust swings, the state went back into drought for at least three years beginning in water year 2018–2019.

6.3.1   Miracle March

In the beginning of 1991, California was in the midst of its worst drought since the Dust Bowl days in the 1930s. Lake Tahoe was at its lowest level since measurements began and snow amounts were at record lows. Ski areas were experiencing economic hardship as retail sales and lodging were down by a large amount and unemployment had more than doubled.

The snow started on March 1 and continued until Lake Tahoe accumulated 50 inches. Skiers were happy. But it didn't stop there. Some ski resorts received about 20 feet of snow, nearly half their annual average. Lake Tahoe's snowpack quadrupled.

Another *Miracle March* occurred in 2011, a near record-setting month for the state with 205% of the long-term average rainfall.

6.3.2 Rainfall Records

California holds several precipitation records, nationally and globally. A world five-minute rainfall record occurred on February 5, 1976, at Haynes Camp with 2½ inches of rain. The 80-min rainfall record in the United States took place on Aug. 8, 1891, with a cloudburst over Campo, California (40 miles southeast of San Diego). Another conterminous United States record for the highest rainfall in a calendar month was 71.54 inches at Helen Mine in January 1909. However, these records have been shattered by our tropical state, Hawaii, so these records apply only to the conterminous United States. The mostly west–east oriented San Gabriel mountains, along with their eastern extension, the San Bernardino Mountains are very efficient at blocking Pacific storm moisture from the lee sides of the mountains. This moisture, however, can produce some of the highest rainfall totals in the United States on the mountains' windward side, comparable to that produced by Atlantic hurricanes. Atmospheric Rivers are responsible for almost all (92%) the West Coast's heaviest three-day rain events, according to Martin Ralph of the University of California, San Diego (UCSD) Center for Western Water and Weather Extremes.

While these high totals occurred in mountainous locations, the lowest annual average in the United States, 1.66 inches, is from Greenland Ranch, Death Valley, in the lee of the Sierras.

See Snowfall below for snow records.

6.4 Floods

With too much rain individual streams and rivers overflow their bank causing floods. Several types of large-scale weather systems, such as atmospheric rivers and cut-off lows, often may last several days, leading to flooding of large and small streams.

Periodic floods, sometimes with major impacts occur throughout California's 58 counties with every county having had at least one flood in the past 25 years. Floods do have some beneficial effects. They control erosion and sedimentation, replenish soils and beaches, recharge groundwater, and support many river and coastal floodplain habitats for California's endangered species. When floods do occur in populated areas, they threaten life and property and often cause economic ruin. Like other parts of the United States, a leading risk factor for California floods is situating many large cities along major rivers. Floods have occurred on all the major rivers, from the north to south. They occur in all seasons and in all regions. Mountainous steep slopes can lead to mudflows, while flash floods in desert washes can cause sudden disasters. Flood risk varies across the state, generally increasing with storm frequency and intensity, along with urban development in floodplains and unstable hill and mountain slopes. Climate change may lead to more variability in meteorological and hydrologic conditions with increasingly intense floods, along with increasingly severe droughts (Fig. 6.8).

The most devastating California flood ever recorded occurred in the winter of 1861–1862. There were record floods throughout the state in 43 days of persistent storms. During December and January, areas received 200–400% of their average rainfall from numerous atmospheric rivers and cold polar storms. As a result, the Central Valley became a large inland lake. Sacramento, the capitol, was inundated by several feet of water. Governor Leland

Fig. 6.8 1861–62 Flood was unprecedented in the state's history. A similar flood today would be the costliest disaster in California history (United States Geological Survey 2018)

Stanford inauguration took place with boat transports. The flood was considered a once-in-a-30,000-year event. Geological evidence indicates that similar massive California floods happened six times between the years 200 CE and 1700 CE according to a government report. The consequences of such a megaflood like 1861–2 today are described by the USGS study, ARKStorm, a Hollywood-like disaster scenario that claims damages could run into the hundreds of billions of dollars.

Major floods have changed the way California deals with its water resources and with reducing flooding. Two examples highlight the many events that have shaped the state's flood control.

6.4.1 March 1907 and December 1909 Floods in Sacramento

After early twentieth-century flooding of the rivers near Sacramento, new methods of flood control were implemented. The idea of using levees was abandoned and bypasses, and overflow weirs were implemented to control the Sacramento rivers from flooding.

6.4.2 1938 Los Angeles Flood

A series of storms that arrived in southern California on February 28, 1938, and lasted five days left behind a messy disaster in Los Angeles, Orange, and Riverside counties as well as adjacent parts of the region. Rivers overflowed their banks and over 100 people were killed. There was also widespread damage and destruction of homes and businesses This particularly damaging storm followed a series of similar disasters that led the city to dig a deep, wide concrete flood control system otherwise known as the Los Angeles River. (See Appendix D for other notable winter storms.)

6.4.3 Flood Frequency

Geologic evidence shows that massive floods occur every 200–300 years in California. These are probably due to atmospheric rivers. Estimates on how frequent major rainstorms occur also give insight into flood risks throughout the state. Based on previous rainfall records, hydrological surveys can estimate how much precipitation would fall in an hour, day or longer once in a decade, century, or millennium. For example, a 100-year flood would have a one-in-one hundred chance of happening in any one year. For much of California west of the Sierras, an expected 24-h rainfall amount typically increases with elevation. Downtown Los Angeles would have a 7-inch storm in 24-h once a century and 10-inch storm once in 1000 years. In higher Pasadena, the values increase by about 40% for the same periods. In the mountains above Los Angeles, Big Bear Lake's values are 12 inches for a one-in-100-year event and 21 inches in one day for a millennial event. On the lee of the Sierras, values generally decrease to about 50% of the Los Angeles value. These rare events would certainly create chaos.

The Indio storm, a desert thunderstorm derived from a tropical storm on September 24, 1939, dropped 6.45 inches of rain. One would not expect a storm of that magnitude to return for 55,000 years!

6.5   Mudslides and Landslides

Associated with heavy rainfall, especially in steep terrain, are landslides and mudslides. Although not all heavy rains produce landslides, the extra weight of saturated soils is a major player in these hazards. But there is a combination of factors that contribute as well: intensity, temporal spacing, rainfall duration, slope steepness, geology, vegetation, construction, previous precipitation, and preceding wildfires. What encourages these hazards are the seasons-the hot, dry summer, the fall/early winter Santa Ana winds and fires and then followed by winter storms.

On January 4, 1982, a fast-moving frontal cyclone tracked through the San Francisco area. It affected areas from south of San Francisco northward to the San Francisco Bay area. Heavy rains were associated with this storm and over a 36-h period much of the area had received half its annual rainfall. On January 4 and 5, some places received as much as 24 inches of rain. On the second day of the storm, landslides began on many of the hills surrounding San Francisco Bay. Falling trees accompanied the landslides and nearly 8,000 homes and businesses were damaged. A landslide even shut down the Golden Gate Bridge. Fifty-three deaths were confirmed with damages exceeding $700 million in today's dollars Most of the damage was from a two-mile-long mudslide that hit Santa Cruz County, the worst storm to hit the region in a quarter century. This mass of water originated near the Hawaiian Islands, a Pineapple Express or atmospheric river. However, this storm occurred neither during an El Niño nor La Niña event, but rather neutral conditions. This event, the worst natural disaster since the 1906 earthquake, resulted in the region being declared a federal disaster area.

6.5.1 El Niño Flooding, Mudslides

Strong El Nino events like 1982–83 and 1997–98 often cause extensive flooding and mudslides. Heavy rainfall on February 2–3, 1998, led to widespread landslides in the San Francisco Bay area. A number of homes and properties were damaged by slides. Damages from mudslides were greater than a damaging 1982 storm due to prolonged rainfall throughout 1997–98 as opposed to one great storm in 1982.

February 1998 was one of the wettest Februarys on record for the state. Heavy rains resulted in dozens of major landslides, some deadly. Several homes in Laguna Beach were destroyed by landslides on Feb. 24, 1998, killing two people.

6.5.2 1997 New Year's Yosemite Flood and Landslides

December 26, 1996, through January 3, 1997, saw a series of storms that affected northern and central California. Storm intensity peaked on New Year's Day flooding Yosemite Valley. The last time this happened was more than a century ago. Flooded rivers and landslides followed. Many roads were impassable. Eight deaths were reported, and most of northern and central California was declared a federal disaster area.

6.5.3 Fires, Rain, and Slides

Often mudslides occur after fires remove vegetation from hillsides leaving soil and debris to move downslope by gravity when heavy rains add weight to unstable slopes. After the largest wildfires in California history, the Thomas Fire near Santa Barbara's Montecito area, in the previous month, heavy rains on January 8–9, 2018, washed away soil and debris destroying 65 homes and leading to 21 deaths. Damages ran into the tens of millions of dollars as homes in the area had an average value of over $4 million. Heavy debris flow including mud and boulders flowed down into the town of Montecito (near Santa Barbara) from the Santa Ynez mountains. Many people and businesses lost power. In places, the mud and debris flow reached the beaches. A stretch of US Highway 101 was closed between Santa Barbara and Ventura.

With the fire season now expanded almost throughout the year, early winter storms are watched carefully for slides in burn areas (Fig. 6.9).

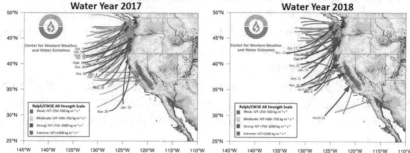

Fig. 6.9 Atmospheric rivers were more active in California in 2016–17 (water year 2017) than 2017–18 (water year 2018) resulting in a wet year and heavy snowfall. See also Figs. 6.11 and 6.12 (University of California San Diego)

6.6 Snow

Snow is the most common form of solid precipitation in California, although hail and sleet also occur. Although snow has occurred in most California locations, it is rare in coastal areas (except the coastal mountains) and the Great Valley. Traverse Range and the Cascades. Sierra Nevada translates in Spanish to "snow covered mountains." In the Sierra Nevada, snowfalls as low as 2000 feet are common, but large amounts occur above 4000 feet and increase to 7000 or 8000 feet. Mountain highways can be shut down for periods of hours to days from drifting snow. East of the Sierra Nevada snow amounts at elevations of 4000 feet, or higher, are usually quite small.

Pacific winter storms that bring rain to lower elevations in California can produce huge amounts of snow in the mountains, measured in feet. California mountain snow has relatively high water content. Referred to as "Sierra Cement," these wet snows differ from more powdery snows further inland.

Winter snow is critically important to the state's water reserves, to supply the state's needs through the hot, dry summer. But it is not just the amount of snowfall that is important to the state's water supply. Three important aspects of snow are the following (Los Angeles Department of Water and Power):

- *Cumulative Precipitation*—this is the overall accumulation of all forms of precipitation including rain and snow. This variable tells how "wet" a winter is.
- *Cumulative Snowfall*—this is the total accumulation of only the snow that falls during a water year. The majority of precipitation that falls during the winter at higher elevations is in the form of snow. However, measurement of the amount of snow differs from actual precipitation due to the variable water content of snow under different conditions and the amount of rain that falls. For example, snow that crystalizes at very low temperatures has a low water content.
- *Maximum Snowpack Depth*—this variable measures the depth of the snow on the ground. This is the parameter that determines what the "base" is at ski resorts and how much snow is covering the ground.

Higher temperatures may lead to snowmelt, sublimation, or even rain instead of snow. Atmospheric rivers may bring lots of moisture but being from tropical latitudes may not add as much snow as storms from polar regions. Snowpack that stays late into the spring is more valuable as a resource into the hot, dry summer months than large amounts of early snow that runs off early, causing floods.

In 1982–83, during the strongest El Niño of the Century, the Sierra Nevada collected 747 inches of snow at Echo Summit. The same location recorded the second-greatest United States 24-h snowfall record: 67 inches (5.6 feet) on January 4–5, 1982. The station sits at an elevation 7,450 feet, approximately ten miles southwest of South Lake Tahoe, on the upper western slopes of the Sierra Nevada.

6.6.1 Snow Records

The snowiest parts of the Sierra Nevada receive 250–500 inches of snow a year. Blue Canyon, California, ranks as the snowiest location in the lower 48 (with reliable records). At nearly 5000 feet, it gets 254 inches a year on average. Some places at higher elevations may average 500 inches of snow a year, but data are less reliable.

6.6.2   Tamarack, California

Tamarack is at an elevation of 6913 feet on the west slope of the central Sierra Nevada near Yosemite National Park. Tamarac holds the United States and world snowfall record for one month, 390 inches in January 1911. Tamarack also holds the United States snowfall record during one season, 884 inches in 1906–07. United States seasonal snow depth record: 454 inches set here on March 10, 1911, is the most snow on the ground at one time.

The greatest measured 24-h snowfall in the United States, 60 inches occurred on the western slopes of the Sierra Nevada at Giant Forest, Sequoia National Park on January 18–19, 1933. Farther to the north, in February 1959, the greatest single continuous snowstorm in the world dumped 189 inches (nearly 15 feet) of snow at Mount Shasta Ski Bowl, with drifts burying entire ski lifts. A total of 236 inches (19.7 feet) of snow fell at Mt. Shasta that February, another record for the month.

6.6.3   Donner Summit, California (Donner Pass)

Donner Pass is an important Northern Sierra Nevada Mountain pass and is just over a mile from I-80 and close to Truckee, CA, at an elevation of 7399 ft. The first wagon train to reach California crossed Donner Summit followed by many more. Donner Summit has many notable geographic and historic landmarks. It was where the first transcontinental railroad, the first transcontinental highway (the Lincoln Highway), the first transcontinental air route, and the first transcontinental telephone line were situated. It also has an infamous snowy past (Fig. 6.10).

The Donner Party (sometimes called the Donner–Reed Party) was pioneers who set out for California in a wagon train. Delayed and trapped by heavy snow, they spent the winter of 1846–47 in the Sierra Nevada for nearly five months. Reports of cannibalism were later reported due to starvation. Almost half of the party of 87 died in the mountains during that awful winter.

Donner Summit at elevation 7017 ft. broke the US record for the snowiest April with 298 inches (24.8 ft.) in 1880. Four times since 1880 total seasonal snowfall at Donner Summit has exceeded 775 inches, topping 800 inches in both 1938 and 1952.

On April 1, 1880, a strong Pacific storm hit the Sierra Nevada west slope at Cisco Grove with four feet of snow in 24 h. A large snowslide near Emigrant Gap buried Central Pacific Railroad's tracks under 75 feet of snow, ice, and rock. Several more storms brought a record 298 inches of snow to Donner Summit. Avalanches blocked the CP train route for days. Another intense

Fig. 6.10 Donner Pass. This is the site of one of the most infamous chapters in California history in the winter of 1846. Here the Donner Party of pioneers was tragically snowbound for nearly five months near what is now called Donner Lake, shown in the background (Wikipedia Commons Photo "Donner Pass" by Joe Parks is licensed under CC BY-NC 2.0., Creative Commons Attribution-ShareAlike 2.0 Generic License)

storm buried the Summit region on April 20 and 21. It was described as "the heaviest and most protracted one ever encountered on the line of the Central Pacific." Truckee, California nearby was buried under 16 feet of snow, while ice was measured at 10 feet thick on Donner Lake. Donner Summit received almost 67 feet of snow that winter with more than one-third of it falling in April.

6.6.4   2016–2017—The "Wettest" Year on Record-(Snowfall)

Just as rainfall is quite variable from year to year, so is snowfall. Following a five-year drought in which snow amounts were well below normal, 2016–17 had a record amount of measured snow in the central Sierra Nevada, surpassing the El Niño 1982–83 record year. This occurred two years after the lowest snowpack recorded in the region (Fig. 6.11).

A storm that meteorologists were calling "the biggest of the season" buried the Sierra Nevada with snowfall across five days, February 28–March 4. The five-day storm totals were close to seven feet in some areas above 7500 feet. By March 7, 2017, snow conditions in the central Sierras were well above the previous best year (1982–83) with 253% of normal snowfall totals.

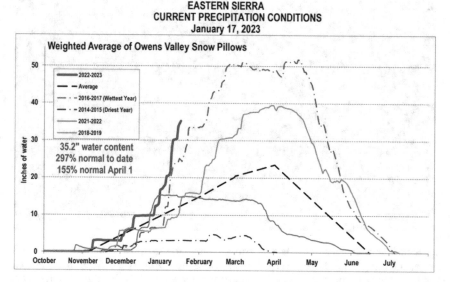

Fig. 6.11 Accumulated precipitation for 2023 for the eastern Sierras, with the wettest and driest years compared to the normal (Los Angeles Department of Water and Power)

The following year's total dropped to 65% of normal for the central Sierras by April 1. However, wettest does not mean the snowiest. As previously mentioned, warm, wet snow produces more water equivalence than cold, dry snow of the same depth. The snowfall in 2022–23 has surpassed the 2016–17 amounts in early January.

6.6.5   Snow in the Big Cities

Snow has been seen in just about every location in California. Above elevations of 2000 feet, snow may occur and above 4000 feet, it is the predominant winter precipitation. For example, located at about 6200 feet, the city of South Lake Tahoe's normal snowfall is over 140 inches. However, snow in California's coastal cities is quite rare.

Saturday, January 20, 1962, had been a chilly, but not particularly unusual, winter day in San Francisco with high temperatures reaching the upper 40 s. Early the next morning, however, things had changed. Long-time residents woke to a surprise. Beth, a high school student, was awakened by a phone call at 6 a.m. It was a friend inviting her to go sledding! Looking out her window, she was amazed to see everything covered with a blanket of snow. Anecdotal reports revealed that about three inches had fallen in San Francisco and parts of the San Francisco Peninsula. At the airport, snow had begun around 4 a.m. and continued to nearly 9 a.m. with 1.5 inches officially recorded. This was very rare–the last time measurable snow had fallen in San Francisco was 1887!

Snowfall is rare along the coast, though the probability increases as one travels towards the Oregon border. San Francisco has recorded just a few snow days, but because of the hilly terrain, chances of a white cover are more common at higher elevations. San Francisco's financial district received an inch of snow was December 11, 1932. On February 5, 1976, a cold Canadian air mass produced an inch of snow in the city. Since then, only higher elevations like Twin Peaks got any white stuff, as it did in February 2011.

In San Francisco, during the nineteenth century, measurable snow occurred about once per decade. Notable was the two-and-a-half-inch fall that occurred on December 30, 1856. Temperatures remained above freezing with 36 °F being the lowest temperature recorded during the event. Another significant snowfall occurred on January 12, 1868. The *Alta California* reported "Snow covered all the roofs and sidewalks even in the lower part of town, and the hills were for the first time in years completely whitened." The 1880s was a decade with numerous winter storms in the United States. During this period, San Francisco reported snow in four winter seasons. Including a steady snow that accumulated to 3.5 inches in the downtown area on New Year's Eve, 1882. During the twentieth century, however, it became even rarer, and since 1962, only traces of snow have been recorded at San Francisco Airport. The snow of January 20, 1962, was the heaviest and most widespread in over 80 years Snow was reported again on February 5, 1976, in the San Francisco Bay area, but accumulations were generally limited to higher elevations with only a trace of snow recorded at the airport, although there were reports of up to one inch downtown.

Sacramento is about 90 miles east of San Francisco. It also experienced snow on the two days mentioned. In January 1962, Sacramento only recorded a trace, but in February 1976, two inches were reported.

With an average January temperature of 57 °F, snow in downtown Los Angeles and at Los Angeles International Airport is even rarer than in San Francisco. Although news reports indicate snow fell six times in the twentieth century, official reporting stations indicate less than that. Many of the events occurred only at higher elevations in the city, like the Hollywood Hills and San Fernando Valley. However, on the night of January 10–11, 1949, occasional light snow was reported at Los Angeles International Airport. Because temperatures remained several degrees above freezing, no accumulations were reported. It is likely, that higher elevations in the city did receive measurable amounts. Although no official observations are available, credible news reports indicate that about two inches of snow fell on downtown Los Angeles on January 15–16, 1932, the highest amount ever reported.

The January 1949 snowfall event was heavily reported by the media with photos of Glendale College students hurling snowballs at each other. The last snow recorded in downtown Los Angeles was during a 1962 storm, though only a trace. However, the San Fernando Valley had snow in 1957, again in February 1989 and in 2007 along with some at Malibu along the coast.

Though rare, snow was recorded in San Diego only five times in over 125 years of record-keeping. Snow flurries were last seen in San Diego on February 14, 2008, at around 1700 to 1800 feet, and the last measurable snowfall to hit various higher parts of the city took place on December 13, 1967.

6.6.6 California Skiing

There are several skiing venues in both northern and southern mountains of the state. In southern California, skiers have a short drive into the San Bernardino Mountains to reach resorts like Bear Mountain, Snow Summit, and Mountain High. Farther north, the Sierra Nevada range is the most popular ski destination in California. Mammoth Mountain, in the Eastern Sierra, is popular for reliable snows. Within the Sierra Nevada, the Lake Tahoe region has the most concentrated amount of ski areas in all of California. On the lake's north side, Squaw Valley's slopes and cliffs provide even extreme skiing, while Sugar Bowl receives enough snow to satisfy downhill crazies. On the south side, Heavenly and Kirkwood provide a broad range of advanced skiing and snowboarding terrain. Six miles south of volcanic Mount Shasta (14,162 feet) in the Cascades, Mount Shasta Ski Park benefits from 275 inches of annual snowfall.

The 2016/2017 ski season produced record-breaking snowfall totals. They called it *Snowmagedeon*. Many resorts extended their seasons, with five resorts still open in summer. Squaw Valley, California, was open past July 4 for the first time in their history. Several ski areas were in the United States. top 10 for 2016–17 snowfall totals including Sugar Bowl Resort with 795 inches (second only to Mount Baker's 866 inches). Mammoth Mountain set a January record for snowfall as totals reached 246 inches from the start of the month to January 24 (Fig. 6.11). The previous record was a *measly* 209 inches (Fig. 6.12).

Precipitation through Jan 26, 2017 compared to normal January precipitation

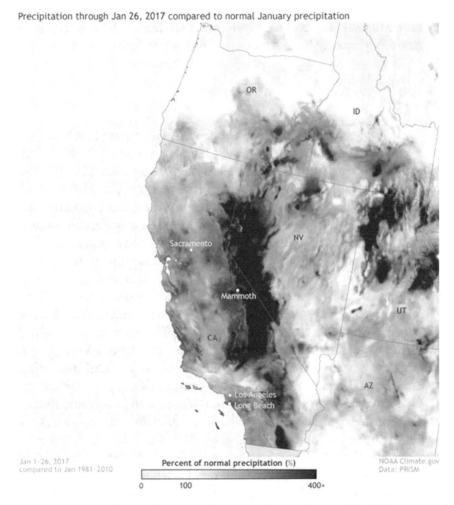

Fig. 6.12 Percent of normal precipitation totals through January 26, 2017, compared to normal January precipitation. Atmospheric rivers during the month contributed greatly to these totals (PRISM, Oregon State University, data from NOAA)

Recreation, such as skiing, brings with it jobs and supports local economies. When there is less snow fewer skiers head for the slopes. California is the most sensitive to snow fluctuation. In a good season, high snow levels could add several hundred million to the economy. A poor season, however, cuts similar amounts. Other winter sports are also affected, such as cross-country skiing and families from the cities heading to the mountains to toboggan or throw snowballs skip these activities during low snow or drought years.

When temperatures are cold enough, snowmakers create their own snowstorms, particularly in southern California mountains. Global warming and frequent droughts may be hazardous to the skiing industry.

6.7   Drought

California has had a long history of drought. In fact, most years have below normal precipitation. But the frequency, duration and intensity of droughts in the state may be on the rise (see climate change Chap. 11). The latest drought to hit the state, since the turn of the century, has been called a megadrought. What's a megadrought? It is when there is a long duration of drought (dry spell) lasting a decade or longer. Such a megadrought occurred in the West in the mid-thirteenth century leading to migration of Native Anasazi in the southwest and causing suffering to California tribes.

There are different types of droughts in California: permanent, seasonal, and occasional. For permanent think of the Sahara or Mohave deserts. Deserts lack water nearly every year and cannot sustain vegetation. The Mediterranean summer dry climate is seasonal drought. We do not expect much rain during the warm summer months. While these two types of droughts are expected, the third is not a regular feature. And it is this kind that impacts the state's natural resources. And yet drought is not simply the lack of rainfall. One must consider duration and intensity as well as soil moisture and temperature, even humidity and winds. Often droughts are classified as agricultural, hydrological, and meteorological. Each focus on a different aspect of the water budget. One considers water needs of crops, the other may deal with streamflow or lake levels. The third may be defined in terms of precipitation deficits, or how much below normal is precipitation over a given period of time.

The United States Drought Monitor (Fig. 6.13) is updated weekly showing a map of drought conditions using 5 categories of drought, from abnormally dry, D0, to exceptional, D4.

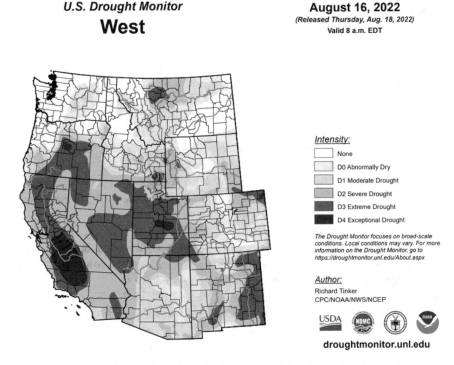

Fig. 6.13 US Drought Monitor for mid-August 2022 (US Drought Monitor)

More on the drought that persisted through the fall of 2022 is covered in Chap. 11 (climate change) and 13 (agriculture). Atmospheric rivers late in 2022 into 2023 greatly improved the drought situation throughout the state.

6.8 Precipitation Trends

From 1950 to 2009, there has been a modest increase in heavy precipitation events in southern and central California, but a decline in heavy precipitation events in northern California. On the other hand, studies found that for the period 1925 until 2007, stations in central and northern California had increasing annual precipitation, while southern California showed either no change or some decline.

Bibliography

Aldrich, J. H., & Meadows, M. (1956). *Southland Weather Handbook*. The Weather Press.

Anderson, B. R. (1975). *Weather in the West*. American West Publishing Co.

Berkeley Library, University of California. (2022, December 21). https://guides.lib. berkeley.edu/bancroft-library

Bora, K. (2014, December 6). California drought of 2012–2014 is the worst in 1,200 years: Study. *International Business Times*. https://www.ibtimes.com/cal ifornia-drought-2012-2014-worst-1200-years-study-1737522#:~:text=Reuters% 2FNoah%20Berger%20Record%20high%20temperatures%20and%20very% 20low,Minnesota%20and%20Woods%20Hole%20Oceanographic%20Institu tion%20in%20Massachusetts

California Department of Water Resources. (2015, Februaary 2). *California's most significant droughts: Comparing historical and recent conditions*. Sacramento: Dept. of Water Resources. https://cawaterlibrary.net/document/californias-most-signif icant-droughts-comparing-historical-and-recent-conditions/

California Department of Water Resources. (2022, December 5). *Flood*. Sacramento: Dept. of Water Resources. https://water.ca.gov/Water-Basics/Flood

Cayan, D. R., Dettinger, M. D., Diaz, H. F., & Graham, N. E. (1998). Decadal variability of precipitation over western North America. *Journal of Climate, 11*, 3148–3166.

Feinstein, L., Phurisamban, R., Ford, A., Tyler, C., & Crawford, A. (2017). *Drought and inequity in California*. Pacific Institute. https://pacinst.org/wp-content/upl oads/2017/01/PI_DroughtAndEquityInCA_Jan_2017.pdf

Fierro, A. O. (2014). Relationships between California rainfall variability and large-scale climate drivers. *International Journal of Climatology, 34*(13), 3626–3640.

Fritts, H. C., & Gordon, G. A. (1980). *Annual precipitation for California since 1600 reconstructed from western North American tree rings*. California Department of Water Resources.

Hall, A. (2016). *How California climate shapes water resources*. http://workshop.cal tech.edu/caH2O/presentations/Alex_Hall.pdf

Hereford, R., Ebb, R. H., & Longpre, C. I. (2006). Precipitation history and ecosystem response to multidecadal precipitation variability in the Mojave Desert region, 1893–2001. *Journal of Arid Environments, 67*(supplement), 13–34. https://www.sciencedirect.com/science/article/abs/pii/S0140196306002965? via%3Dihub

Historic California Floods in Photos. (2020, November 5). https://www.nbclosang eles.com/news/california-wildfires/historic-california-floods-storms-rain-images/ 10767/

Ingram, L. (2013, January 1). California megaflood: Lessons from a forgotten catastrophe. *Newsweek*. https://www.scientificamerican.com/article/atmospheric-rivers-california-megaflood-lessons-from-forgotten-catastrophe/

Jong, B.-T., Ting, M., & Seager, R. (2016). El Niño's impact on California's precipitation. *Environmental Research Letters 11*, 054021. https://iopscience.iop.org/article/10.1088/1748-9326/11/5/054021/ampdf

Killam, D., Bui, A., LaDochy, S., Ramirez, P., Patzert, W., & Willis, J. (2014). California getting wetter to the north, drier to the south: Natural variability or climate change? *Climate, 2*, 168–180. https://doi.org/10.3390/cli2030168

Landin, M. G., & Bosart, L. F. (1989). The diurnal variation of precipitation in California and Nevada. *Monthly Weather Review, 117*, 1801–1816.

Lloyd, J. (2021, October 25). *Historical California winter storms: Devastating floods, Landslides, Dust storms.* https://www.nbclosangeles.com/weather-news/historic-california-storms-flooding-rainfall-damage-winter/19208/

Meares, H. (2017, December 21). *LA's snow days.* Curbed Los Angeles. https://la.curbed.com/2017/12/21/16794092/snow-los-angeles-photos

Mitchell, T. P., & Blier, W. (1997). The variability of wintertime precipitation in the region of California. *Journal of Climate, 10*, 2261–2276.

Monteverde, J., & Null, J. (1998). *El Niño and California precipitation.* Western Region Technical Attachment Region Tech. Attachment No. 97-37. https://www.weather.gov/media/wrh/online_publications/TAs/ta9737.pdf

NASA. (2016, March 2). *Atmospheric river storms can reduce Sierra snow.* https://www.nasa.gov/feature/jpl/study-atmospheric-river-storms-can-reduce-sierra-snow

Null, J. (2017). California's stressed water system: A primer. *Weatherwise, 70*(1), 12–19.

Null, J. (2018). *El Niño and California precipitation.* https://www.ggweather.com/ca_enso/ca_elnino.html

Null, J., & Mogil, M. (2010). The weather and climate of California. *Weatherwise, 63*(2), 16–23.

Reports to the Nation. (2022, December 5). *The North American monsoon.* https://www.cpc.ncep.noaa.gov/products/outreach/Report-to-the-Nation-Monsoon_aug04.pdf

Rowe, P. (2007, December 13). *The day it snowed on San Diego.* San Diego Union-Tribune.

Rutz, J. J., Steenburgh, J., & Ralph, F. M. (2014). Climatological characteristics of atmospheric rivers and their inland penetration over the western United States. *Monthly Weather Review, 142*, 905–921.

San Francisco Snowstorms. (2022, December 5). http://thestormking.com/Sierra_Stories/San_Francisco_Snowstorms/san_francisco_snowstorms.html

Snow Brains. (2018, March 13). *"Miracle March 1991" The March that saved Lake Tahoe.* https://snowbrains.com/miracle-march-1991-march-saved-tahoe/

Sumner, T. (2015). California drought worst in at least 1200 years. *Science News, 187*, 16.

United States Geological Survey. (2018, January 23). *ARkStorm scenario.* https://www.usgs.gov/programs/science-application-for-risk-reduction/science/arkstorm-scenario?qt-science_center_objects=0#qt-science_center_objects

United States Geological Survey. (2022, December 5). *2012–2016 California drought: Historical perspectives.* https://ca.water.usgs.gov/california-drought/california-drought-comparisons.html

United States Geological Survey. (2011, February 4). *ARkstorm: California's other big one.* https://www.usgs.gov/centers/pcmsc/news/arkstorm-californias-other-big-one

University of California Riverside. (2022, December 21). *California digital newspaper collection.* https://cdnc.ucr.edu/?a=cl&cl=CL1&sp=DAC

Winkler, J. A. (1992). Regional patterns of the diurnal properties of heavy hourly precipitation. *The Professional Geographer, 44*, 127–146.

7

Fog: A Menace or Friend

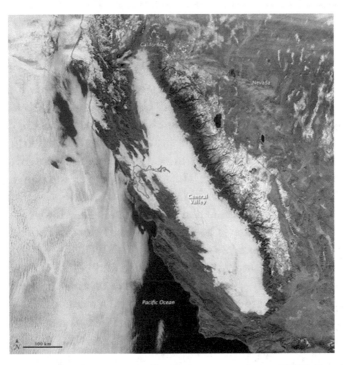

Fig. 7.1 Tule fog extends over 400 miles during most winters (NASA Terra), leading to poor visibilities and occasional chain-reaction crashes

© The Author(s), under exclusive license to Springer Nature
Switzerland AG 2023
S. LaDochy and M. Witiw, *Fire and Rain*,
https://doi.org/10.1007/978-3-031-32273-0_7

Fig. 7.2 Advection fog through the Golden Gate (NOAA 2022)

Saturday morning, November 7, 2007, was a normal fall morning for central California. Thick fog, not unusual for that time of year, shrouded much of the area, including the area's highways. Unfortunately, that day was to be marred by a catastrophic event. At about 8 a.m., a horrendous chain-reaction crash occurred on State Highway 99 near Fresno in visibilities estimated to be less than 600 feet. News reports indicate 108 vehicles were involved with two deaths and 41 others injured. The fog was a type of radiation fog, locally called Tule fog after the Tule plants that prevail in California's Central Valley. Tule fog can sometimes affect California's Central Valley for weeks at a time during late fall and winter. The satellite image below from NASA clearly shows how extensive this type of fog can be (Fig. 7.1).

The geography of California from its rugged and beautiful coastline to its coastal hills, inland valley, and finally the Sierra Nevada makes it prone to several different types of fog. By definition, fog is a cloud attached to Earth's surface where visibility is less than 1000 m or 5/8 of a mile. (By definition, similar conditions with greater visibilities are called mist.) One type, marine fog (a type of advection fog) predominates during summer and, at times, occurs along almost the entire coastline. Many pictures showing the Golden Gate Bridge shrouded in this type of fog are often seen on postcards. Figure 7.2 is a picture of advection fog. To the north, Point Reyes is one of the foggiest places in the United States with over 200 foggy days per year.

Types of fog that occur mainly during the cool season include radiation fog, upslope fog and frontal fog. So, how does fog form and why is California subject to so many different types of fog?

7.1 How Does Fog Form?

Water is present in Earth's atmosphere in liquid, frozen, and vapor states. We can see water when it is either in the frozen or liquid state. This includes clouds which are composed of tiny droplets of water and/or small crystals of ice. Water vapor, like the other gases in the atmosphere (oxygen and nitrogen, for example), is invisible.

As the atmosphere warms, its ability to hold moisture in the form of water vapor increases. Temperature measures how warm the air is and how much moisture it can hold. Dew point is a measure of how much moisture is in the air. When these variables are measured, they are usually reported like this: 63/52. This would indicate a temperature of 63 and a dew point of 52.

Relative humidity tells us how much water vapor is in the air compared to how much the air can hold at that temperature. It can be calculated by knowing both the temperature and dew point. The bigger the difference between temperature and dew point, the lower the relative humidity is. For example, if the temperature were 100 and the dew point 52, the relative humidity would be 20%. If the temperature lowers to 80°, but the dew point remains the same, the relative humidity is now 38%. Further drops in temperature will increase the humidity even more. When the temperature reaches 52° (same as the dew point), the humidity is now 100%. If the temperature goes down just a little bit more, the air can no longer hold all the moisture in the form of vapor. If this occurs near Earth's surface moisture will condense and, depending on conditions, will form either fog, dew, or frost (at temperatures less than 32 °F).

7.2 Advection Fog

If you have ever driven along any California highway near the ocean, you have probably experienced advection fog. It can come inland any time of day, but late afternoon during summer is a common time. Advection fog comes from the chilling of the air above a cool ocean current. During the warm season, California is dominated by the North Pacific High offshore and low pressure in the interior valleys (Fig. 7.3)

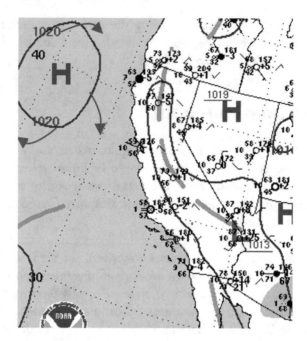

Fig. 7.3 Wind flow off the California coast in summer (NOAA 2023 Daily weather maps, surface weather map, graphic by Harrison Blizzard)

To equalize pressure, the air moves from high pressure to low pressure, but because of the Earth's turning, the resultant Coriolis Force causes the air to move more parallel to the coastline. This, in turn, exerts a force on the water, pushing the surface water to the west. This surface water is then replaced by colder water from below in a process called "upwelling." As a result, surface water just to the west of San Francisco can have summertime temperatures as low as 52 °F. Warmer, moist air moving over the cold water is chilled to its dew point and below, condensation occurs, and a deep layer of fog may form. The fog may remain offshore for long periods, but it can move inland suddenly, reducing visibilities along the immediate coast. Further inland, it tends to rise, affecting only the higher elevations as fog, but resulting in a layer of low clouds elsewhere. These low clouds, or stratus clouds, are sometimes referred to locally as "high fog."

7.3   Radiation Fog

Frequently, when we think about radiation, nuclear radiation or the type associated with X-rays comes to mind. Heat radiation is what is discussed here and that is not dangerous. During the day, Earth warms as it absorbs heat from the sun. At night, Earth's surface cools as it radiates heat energy into space.

As the land cools, so does the air in direct contact with it and this cooling effect spreads upward by the process of conduction. If conditions are just right, fog will form. For this to happen, light winds are needed to help mix the cooler air through a thick enough layer of the atmosphere above. If there are no winds, condensation of moisture on Earth's surface is likely. If winds are too strong, cold air is mixed too thoroughly upward and fog will not form. Radiation fog will often dissipate during the day when exposed to the sun's rays.

There are areas in the world where geography traps a layer of cold air and fog forms. The cold air and the fog layer associated with it get deeper each night until a front breaks it up. The Central Valley of California is one of these places. Sometimes, this fog will last for days or even weeks. Usually, conditions will improve slightly in the afternoon, but deteriorate as soon as the sun goes down. Heat from the sun helps dissipate the fog, but as the fog layer gets deeper, the sun can only penetrate the higher layers of the fog. A disturbance, (for example, a cold front) will eventually cause the fog to dissipate.

7.4   Upslope Fog

Unlike radiation fog and advection fog, upslope fog does not affect major population areas or airports in California. It occurs when moist air flows up a mountainside. The air cools at a rate of about 5.5 °F per 1000 feet of altitude. This is called adiabatic cooling and occurs whenever air rises in the atmosphere. When the temperature is cooled to the dew point, saturation occurs. Further cooling may result in fog which can be found in mountainous areas of the state.

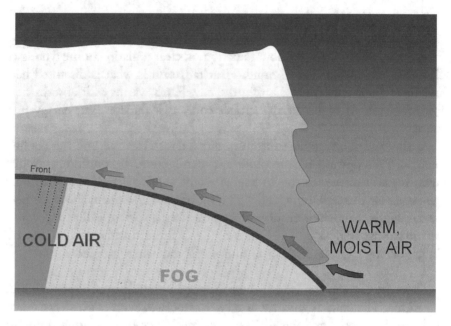

Fig. 7.4 Warm air moving over the top of a cold air mass (Federal Aviation Administration)

7.5 Precipitation Fog

Precipitation fog, also called frontal fog, occurs ahead of warm fronts and to the rear of slow-moving cold fronts. Because warm air is less dense than cold air, it will ride over the top of cold air in a process called over running. As warm air rides over the top of a cold air mass, relatively warm rain falls into the cold air below. Since cold air cannot hold as much moisture as warmer air, any liquid water that evaporates quickly saturates the air, with fog resulting, as shown in Fig. 7.4. It can sometimes be hazardous but is usually less intense than many radiation fog events. Unlike radiation fog which has its origin at Earth's surface, precipitation fog works its way down from the base of a precipitating cloud, so it tends to be more intense at higher surface elevations.

7.6 Steam Fog

Another form of fog resulting from evaporation is steam fog. This type of fog, which is rare in California, occurs when cold, dry air moves over warmer water, such as a lake or pond, usually in the fall before the lake has had a chance to cool. Water evaporating into the cold air may lead to saturation near the ground, forming fog-like steam rising off a warm mug of coffee. The rising moisture condenses making for an eerie landscape sometimes used to set the mood in horror films.

Rarely, a variation on this type occurs when rainfall saturates the ground followed by cooling under clear skies at night. This happened along the Interstate-5 in Orange County leading to near-zero visibility and multiple fender-benders.

7.7   Fog and Transportation

As we have already seen, fog can contribute to highway accidents, but it also remains a major hazard for aviation. The amount of fog that occurs in the coastal airports of California is highly dependent on ocean temperatures. In general, in years when ocean temperatures are colder than normal, fog tends to occur with greater frequency. Likewise, when water temperatures are warmer than normal, the amount and frequency of fog tend to decrease.

On Sunday, January 26, 2020, a tragic accident occurred in the Santa Monica Mountains outside of Los Angeles. Star basketball player, Kobe Bryant of the Los Angeles Lakers, along with eight others, including his daughter and the pilot, were killed when the helicopter they were in slammed into a hillside in dense fog at an altitude of a little over 1000 feet above mean sea level (msl). The National Transportation Safety Board (NTSB), in their preliminary report, stated the aircraft had reached an altitude of 2300 feet msl, about 100 feet below the top of the fog layer, before descending and crashing into the mountainside. This accident reminds us that fog demands respect and is often a contributing factor in aviation and other transportation accidents (Fig. 7.5).

7.8   Trends in Fog

Over the last half century, the amount and intensity of fog have decreased throughout California. In some cases, the reasons are evident, but in others they are not clear. For example, dense fog (visibility less than ¼ mile) has been studied extensively at Los Angeles International Airport. The amount has decreased from over 200 h during the winter months in the 1950s to less than 10 h in recent years (Fig. 7.6). In fact, some years have reported no dense fog at all. One of the reasons for this appears to be a major decrease in particulate pollution, down more than 90% in the Los Angeles Basin since

Fig. 7.5 View from a maintenance yard showing the helicopter proceeding west on CA 101 (National Transportation Safety Board)

the 1960s. Also leading to these decreases is the urban heat island effect that does not allow temperatures to cool down to the dew point.

In the Central Valley, although less fog has helped with a reduction in traffic accidents, there are also some negative effects. In that region, fruit production is a major industry. The moisture and cool temperatures that accompany the Tule fog that occurs in the fall and winter are important for the fruit. Reasons thought to be responsible for the decline in Central Valley fog include increased urbanization and its accompanying urban heat island effect (the warming centered in cities that results in lower relative humidity readings) and, possibly, climate change.

In that region, fruit production is a major industry. In the fall and winter the moisture and cool temperatures that accompany Tule fog are important for the production of fruit and nuts. In recent years, production of fruit and nuts like cherries, peaches, pistachios, and almonds has declined.

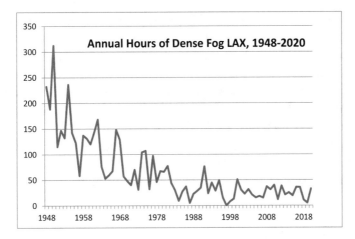

Fig. 7.6 Since the 1950s hours of dense fog at Los Angeles International Airport have plummeted

Another concern is the redwood forests. The redwood forests depend on the moisture from fog during the warmer months, but fog has decreased there also. Early evidence suggests the lack of fog has a negative effect on saplings as well as on mature trees that form the inland edge of the Redwoods. Here, climate change is thought to be the primary culprit.

Nevertheless, despite these trends, parts of the California coast remain among the foggiest places in the United States. For example, Vandenberg Air Force Base, near Point Conception, averages well over 100 days per year when visibility is less than ½ mile.

7.8.1 Mitigation of Fog as a Transportation Hazard

As we saw, fog has many beneficial effects, but it can also cause problems, especially for transportation. So, how can we mitigate these hazards? Lighthouses are probably the earliest attempt, with primitive versions going back millennia to ancient Greece and Egypt. Their main use, however, was to alert mariners of dangerous coastlines, and dense fog made them ineffective. In the mid-nineteenth century, the foghorn was invented. When lighthouses are obscured, foghorns can provide warning to mariners of coastal hazards such as rocks or shoals. Although modern commercial ships typically have their own warning systems such as radar and sonar, foghorns remain useful to the casual sailor. Today, for example, foghorns are located on several islands in San Francisco Bay including Alcatraz, as well as three on the Golden Gate Bridge.

The November 2007 accident (discussed in this chapter's first paragraph) along with other fog-related accidents, prompted the California Department of Transportation to develop and install a fog warning system along a 13-mile stretch of CA 99. This consists of cameras, visibility detectors, and speed detectors. Drivers are warned by changeable message signs along the highway. During foggy conditions, the use of low beams or fog lights, is important as high beams reflect back into a driver's eyes during fog.

Fog may increase the likelihood for an accident and be a contributing factor to it. But while fog presents a challenge, it is ultimately the driver's or pilot's responsibility to adjust to prevailing conditions.

7.9   In Summary

Watching as fog comes in through the Golden Gate is a beautiful sight, but fog on highways or at airports can be treacherous. No matter how you see it though, fog is as an important part of California's complex and varied climate.

Bibliography

Federal Aviation Administration. (2022, December 5). *AC 00–6B - Aviation Weather*. https://www.faa.gov/regulations_policies/advisory_circulars/index.cfm/go/document.information/documentid/1029851

Johnstone, J. A. & Dawson, T. E. (2010, February 16). Climatic context and ecological implications of summer fog decline in the coast redwood region. In *Proceedings of the national academy of sciences of the United States.* https://www.pnas.org/doi/10.1073/pnas.0915062107

LaDochy, S., & Witiw, M. R (2012). The continued reduction in dense fog in the Southern California region: Possible causes. *Pure and Applied Geophysics, 169* (5), 1157–1163. https://doi.org/10.1007/s00024-011-0366-3

LaDochy, S., & Witiw, M. R. (2013). Fog hazard mitigation. In P. Bobrowsky (Ed.), *Encyclopedia of natural hazards* (pp. 338–342). Springer.

LaDochy, S., & Witiw, M. R. (2016). Southern California's fog disappearing act: Climate change, ENSO or PDO? In *Proceedings book of the seventh international conference on fog, fog collection and dew* (pp. 92–95). July, Wroclaw, Poland. http://fog-conf.meteo.uni.wroc.pl/images/publications/FFCD_2016_Poland_proceeding_book.pdf

National Transportation Safety Board. (2023, January 18). *Aircraft accident investigation update.* https://www.ntsb.gov/investigations/documents/dca20ma059-investigative-update.pdf

National Weather Service. (2022, December 5). *Fog over water*. https://www.wea ther.gov/safety/fog-water

Pappas, S. (2014). California Tule fog is becoming increasingly rare (photo). *Live-Science News*, June 5, 2014. https://livescience.com/46121-california-tule-fog. html

Witiw, M. R., & LaDochy, S. (2008). Trends in fog frequencies in the Los Angeles Basin. *Atmospheric Research, 87*, 293–300.

Witiw, M. R., & LaDochy, S. (2015). Cool PDO phase leads to recent rebound in coastal Southern California fog. *Die Erde, Journal of the Geographical Society of Berlin, 146*(4), 232–244. http://www.die-erde.org/index.php/die-erde/article/ view/229/114

8

California Winds and Fire Weather

Fig. 8.1 Pasadena windstorm (Courtesy of Dr. Paula Arvedson)

Don't mind the weather when the wind don't blow.
—Uncle Joe, Buffy St. Marie

S. LaDochy and M. Witiw, *Fire and Rain*,
https://doi.org/10.1007/978-3-031-32273-0_8

The Pasadena, Temple City windstorm occurred on November 30 and December 1, 2011. During this historic windstorm, 5500 trees were damaged with wind speeds reaching 100 miles per hour within the City of Pasadena on November 30 and December 1, 2011. Winds that one would expect with a Category 4 hurricane (140 mph) were recorded on the Sierra Crest Mountain ridge. These winds were reported by the National Weather Service. About 350,000 residents in the San Gabriel Valley lost power, including one of the authors, some for over a week. Flights were diverted from Los Angeles International Airport (LAX) and damage was estimated to be $40 million. Our friends in Pasadena could see stars through what remained of their roof after a sizeable tree dropped onto it (see photo above). Traffic was snarled for days as cars maneuvered around fallen trees and cautiously crossed busy intersections with no traffic lights working.

The storm was analyzed as a mountain wave or downslope wind perpendicular to the San Gabriel Mountains ridgeline, although some local newspapers called it a Santa Ana (Fig. 8.1).

8.1   Overview

Winds are caused by unequal atmospheric pressure often a result of unequal temperatures. The wind then begins to flow from higher to lower pressure. The difference between high pressure and low pressure is called the pressure gradient which is directed from high pressure to low pressure. Larger pressure gradients result in stronger winds.

California is affected by many kinds of wind. Examples include the cool moist onshore flow that affects much of the coast during the warmer months and the dry hot Santa Anas that may bring dry, hot air to the southern California coastal regions While heat and moisture define weather and climate, wind plays an important role as well, since it transports both heat and moisture around our planet. A slight change in wind speed and direction can alter weather, sometimes dramatically, replacing cool marine air with warm, dry continental flow. Most winds result from the uneven heating of the Earth. With its diverse terrain and long shoreline, California presents some unique wind patterns.

In much of California, prevailing winds are from the north or northwest most of the year, because of the influence of the Pacific High-pressure area offshore. However, the mountains and varying terrain across the state can alter these winds. Also, because California is in the zone of the prevailing westerlies at steering levels, during the cool season, frontal systems often move in from the Pacific Ocean, disturbing the typical northwest to north low-level winds.

8.2   Cool Season Storms

As we move into late fall, the storm tracks that have stayed well to the north of California begin to move southward. They continue their southward movement well into winter. Wind direction and speed are modified as high-pressure alternates with low-pressure centers. Often, high pressure over the Great Basin interacts with an approaching low pressure, resulting in strong, sometimes damaging easterly to southeasterly winds in coastal areas. This offshore flow results from the strong pressure gradients directed from the high to the lows. As storms move inland, winds will shift, typically to a southerly or southwesterly direction. Winds may be funneled by terrain resulting in high velocities, especially in mountainous areas.

Winter storms are occasionally accompanied by strong winds that damage property and down power lines. These winds are most often from the south near sea level and from the southwest in the higher mountain areas. Some of the most memorable coastal storms have produced winds that gusted in excess of 100 mph at lower elevations and over 150 mph in the higher reaches of the Sierra Nevada.

On the morning of December 12, 1995, a very deep area of low pressure developed as a "bomb" (a rapidly strengthening low pressure) off the northern California coast. Bringing wind gusts of 60–100 mph and torrential rains, the storm downed thousands of trees in exposed sections of the coast and inland valleys; nearly two million people lost power. That storm tore off the roof of the historic Cliff House in San Francisco and inflicted heavy damage throughout the city.

8.3 Downslope Winds

8.3.1 Santa Ana Winds

Santa Ana winds belong to a type of wind known as katabatic or downslope winds. Other examples are the Chinook which affects the east slope of the Rockies and the Föhn near the European Alps. They occur when there is a combination of strong high pressure over a high plateau and low pressure at a lower altitude. A pressure gradient is directed from the high to low pressure and strong winds result. In the case of the Santa Ana, the high pressure is present over the Great Basin with low pressure along southern California. Winds will then follow the pressure gradient out of the Great Basin into central and southern California. In some locations, as they are funneled through mountain canyons, these winds can be extremely strong and gusty, sometimes exceeding 100 mph particularly near the mouths of canyons. Additionally, as the air descends and compresses due to higher pressure, it warms at about 5.5 °F for every 1000 ft with the relative humidity frequently lowering to less than 10%. With altitudes in the Great Basin generally exceeding 5000 feet, the warming from a Santa Ana can exceed 30 °F. Additionally, with temperatures often reaching near or above 100 °F, the health hazard to humans greatly increases. The Santa Anas normally occur during fall and winter. During both seasons potentially catastrophic winds may occur, however, the Santa Anas that occur in early fall tend to be much more severe, because of the additional hazard of heat and wildfires. Since rain is very rare during the warmer months, the low humidity and high winds can intensify and expand small wildfires. Fire weather watches and high wind watches are issued when conditions are favorable for wildfires. Temporary shutdowns of sections of main highways to campers, trucks, and light cars may also result from strong winds (https://en.wikipedia.org/wiki/Santa_Ana_winds).

During winter, this can result in pleasant temperatures and low humidities in the Los Angeles Basin. Because winter rains have usually fallen by then the fire danger, though not eliminated, is reduced. If winds precede winter rains, wildfires may occur.

The Santa Ana wind is entrenched in American literature. Author Joan Didion called this time "the season of suicide and divorce and prickly dread, wherever the wind blows." Raymond Chandler's short story, *The Red Wind* in 1946, draws a colorful portrait of how hot, dry winds affect human behavior:

There was a desert wind blowing that night. It was one of those hot dry Santa Anas that come down through the mountain passes and curl your hair and make your nerves jump and your skin itch. On nights like that every booze party ends in a fight. Meek little wives feel the edge of the carving knife and study their husbands' necks. Anything can happen. You can even get a full glass of beer at a cocktail lounge. (Chandler, The Red Wind 1946)

The strong offshore winds native to the area near Santa Ana Canyon, south of Los Angeles, were also found in literature in the 1836 novel, *Two Years Before the Mast* by Richard Dana. Dana wrote: "On February 13, 1836, we were called up at midnight to slip for a violent northeaster, for this miserable hole of San Pedro is thought unsafe in almost every wind." The crew had to move the ship southwest of Catalina Island, just off the coast from Los Angeles, to avoid the gale. And during the Mexican War in January 1847, Commodore Robert Stockton, marching through Santa Ana to recapture Los Angeles, reported a strange dust-laden windstorm that occurred while his troops were camped overnight (Figs. 8.2 and 8.3).

8.3.2 California Northers

California Northers occur less frequently than Santa Ana winds and are downslope winds occurring along the west slopes of the Sierra Nevada. As these winds descend from the mountaintops into California's Central Valley, which is near sea level, the air can become very hot, in extreme cases, exceeding 110 °F The name California Northers probably came from the writings of the famous naturalist John Muir, who described them in his book The Mountains of California.

8.3.3 Diablo

The Diablo wind is a hot, dry wind from the northeast that sweeps down the hills east of San Francisco through the canyons of the Diablo Range, running north–south on the east side of the San Francisco Bay. Due to compressional heating of air flowing out of the Great Basin across the Sierra Nevada, these winds result in exceptional heat waves in summer. Mild temperatures will accompany the Diablo in winter, although the strong winds counteract the pleasantness of the temperatures.

Santa Ana winds

In addition to increasing the threat of wildfires, Santa Ana winds can cause trouble for drivers and pilots in Southern California.

1 Desert winds orginate from a clockwise flow of air around a high-pressure system east of the Sierras.

2 Air extends from the mountains, and is compressed and warmed, becoming less humid. This lowers relative humidity and dries out vegetation and can fan any existing fires.

3 Winds squeeze through canyons with gusts between 40 and 60 m.p.h.

4 Strong winds create turbulence for area flights and can make interstate travel difficult.

Source: National Weather Service

Fig. 8.2 Santa Ana winds are a result of pressure gradients leading to downslope winds (National Weather Service)

8.3.4 Santa Lucia and Newhall Winds

Other, similarly caused downslope winds, include the Santa Lucia wind and the Newhall wind. The Santa Lucia wind affects California's central coast. The Newhall wind blows through the Newhall pass down into the San Fernando Valley north of Los Angeles.

8.3.5 Sundowner

Another significant downslope wind in the southern California region is the Sundowner wind in the region of Santa Barbara. The term comes from the fact that these downslope winds typically occur in late afternoon or

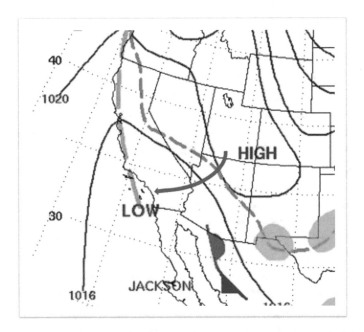

Fig. 8.3 Surface weather map from November 25, 2007, showing typical Santa Ana wind conditions (NOAA, 2023). Graphic by Harrison Blizzard

early evening. Like Santa Anas, a rapid rise in temperature and/or lower humidity accompany Sundowners. The most extreme cases may bring record-breaking high temperatures and sometimes damaging offshore winds. Highs may quickly soar to over 100 °F (37.8 °C). June 17th, 1856, a US Coast Survey engineer recorded an astounding 133 °F (56.1 °C) at 2 p.m. killing animals and fruit off the trees according to the *Coast Pilot of California*. This record was never fully substantiated. Some Sundowners do not extend to the coast, with high temperatures and low humidity inland. Here winds flow across a steep East–West-oriented Santa Ynes Mountains toward the coast due to a strong pressure gradient.

Sundowners can occur throughout the year but are most prevalent in spring. Like Santa Anas, Sundowners are a real threat to spreading wildfires, which has occurred several times. On July 20, 1992, temperature hit 106 °F at Santa Barbara Airport, with 40–60 mph winds in the Santa Ynez Mountains. And on Dec 31, 1995, Sundowner winds blew down trees and power lines along with broken windows on the Santa Barbara wharf.

8.3.6 Washoe Zephyr

Another downslope wind, the Washoe Zephyr, is a moderate westerly wind that occurs mainly in summer on the east lee of the Sierra Nevada. Low pressure forms over the warm high Great Basin while a high pressure forms over the cool central California coastal air. This pressure gradient produces a westerly wind (zephyros in Greek means "west wind") over the Sierras into western Nevada. A zephyr usually lasts several hours during the afternoon and evening when pressure gradients are strongest. At Washoe Lake, between Reno and Carson City, the zephyr is quite reliable with increasing winds in the afternoon and calming down after sunset.

Mark Twain was one of the first to refer to the wind as the Washoe Zephyr when he visited Virginia City, Nevada in 1892. He reported that a Washoe Zephyr "robbed and attacked him." Twain also wrote that the wind was responsible for many bald people in Virginia City. He noted that the dust stirred up by these winds obscured Nevada Territory from sight. The summer afternoon Washoe Zephyr also caused severe dust storms as the area was quite dry year-round (From Mark Twain's **Roughing It**).

8.4 Wind and Fire

Fall is fire season in California, with October being the peak month. Of the most damaging wildfires recorded seven of ten occurred in October. According to Cal Fire, the state's fire monitoring agency, the three deadliest wildfires ever recorded also happened in October. Since it almost never rains in the summer in most of California, conditions are quite dry by the time fall arrives. In most of California, with the possible exception of desert areas with their monsoon rains and occasional light showers in the mountains, summers are dry with very little rain after the month of April. Conditions usually remain dry until November. As a result, by October, the vegetation is parched and tinder-dry in many locations.

While paleo-climatic records from Santa Barbara Channel sediments indicate that frequency of wildfires in the area has not significantly changed in the last 500 years, recent drought, population expansion and the twentieth-century fire suppression policies have made these fires more intense. Rather than the typical fall and early winter fires following the desiccating dry summers, the fire season in California is now year-long. A majority of California large fires involve Santa Ana events. The Cedar Fire, the largest fire in California history at the time (nearly 280,000 acres combined with three other nearby fires accounted for nearly 500,000 acres in late October 2003). Such large fall and early winter fires are often associated with preceding climate patterns (such as ENSO) and recent weather (Santa Ana winds). Rainfall tends to be above normal in the preceding winter adding more fuel to burn. An anomalous high-pressure system over the northwest leads to Santa Ana events. Of the studied fires in southern California, 97% were human-caused and aided by the meteorological conditions. In northern California, lightning-caused fires are of great concern.

In November 2018, Santa Ana-like conditions resulted in the Camp Fire, the deadliest and most destructive fire in California history and the worst in United States history since the 1918 Wisconsin Fire. Butte County, in northern California lost 18,804 structures, killed 85 people, costing some $15–16.5 billion, making it also the costliest fire in state history. On that same day, the Camp Fire started, Santa Ana winds spread the Woolsey and Hill Fires in southern California. In 2018, fires cost over $150 billion in damages, with 3652 deaths in California.

2020 had a record acreage burned by wildfires in the United States and in California. Over four million acres burned, more than twice the previous record of 2 million in 2018. Many of these fires were lightning-caused in August. Droughts over the past decade also contributed to this record fire year. 2021 fire season started faster than the previous year even before the fall Santa Ana winds started but was not as destructive. The 2022 fall was saved by an active monsoon season and early rainstorms.

Fires in the area affected by the Santa Ana occur almost every year and are most likely to turn fearsome when there are Santa Ana winds. The damage to residential property has continued to climb as more people in southern and northern California build expensive homes in the most vulnerable regions of the foothills and mountains. Even if the fires miss communities, there are still costs associated with evacuating residents and suppression of fire. Statewide, wildfire suppression costs can exceed $1 billion annually. The 2010–2020 decade had the most acres burned in history, showing an ominous trend each decade with more and larger fires. Warmer temperatures and more severe droughts have linked climate change as a cause to this upward trend.

During strong Santa Ana events, air traffic is also at risk as strong winds create clear air turbulence and strong wind shear. But there is a benefit. Santa Anas blow pollution out to sea, creates upwelling and nutrients to enhance marine life, and may lead to coastal fog.

8.5   Onshore Summer Wind Flow

In summer, a typical northwest wind flows from the North Pacific High toward the thermal (heat) low-pressure areas located over the Central Valley and the Southeastern Desert area. In the San Francisco Bay area, winds are onshore day and night, with strongest winds usually occurring in the afternoon. As a result, pollution is carried away from parts of the Bay area but tends to be transported further south particularly into the Santa Clara Valley. This wind pattern is not a true sea-breeze. During a true sea-breeze pattern, the wind will reverse direction to offshore at night.

8.5.1   Delta Breeze

Sacramento can tough out its 110-degree days as well as any other sunbaked city. But thanks to the Delta breeze and the onshore breeze from the San Francisco Bay, you can jog after work without risking heat stroke, enjoy a brewski on the patio and sleep with the windows open. The Delta breeze usually doesn't reach Sacramento until evening because it's so far inland. Sometimes it stalls out before reaching because the temperature difference between coast and Valley isn't great enough, or when an inversion layer settles over the region. Sacramento is right at the edge of where the ocean breeze ends.

8.6   Land-Sea Breeze

The sea breeze is important because it impacts fire weather, air pollution, agriculture, and aviation operations, among other things. The sea breeze-land breeze regime is a diurnal shift in winds along the California coast. During warm daytime hours, the land heats faster than the coastal ocean. This temperature difference produces a lower pressure over the warmer, more buoyant air inland and high pressure with the cool, denser air along the coast. The greater the pressure gradient, the faster is the sea breeze. Hence, on hot summer days the sea breeze starts off slowly in early to mid-morning but becomes stronger as the land heats. By mid-day, a brisk breeze flows from ocean inland. By late afternoon, the sea breeze winds peak and extend far inland. In late afternoon and evening the land cools, while the ocean temperatures stay pretty constant. Eventually the air over land becomes cooler than ocean air and the winds reverse direction. The reverse flow at night, called the land breeze, is typically weaker than the sea breeze and on very hot days may not even occur (Fig. 8.4). Sea breezes are strongest in summer and in lower latitudes, while they are weaker in cooler months and at higher latitudes. The cool, marine air interacts with coastal topography, which may block, divert, or funnel these breezes. In the western portion of Los Angeles' San Fernando Valley, sea breezes converge from the Oxnard Plain coming east and the branch from the LA Basin, curling around the Hollywood Hills moving west (Fig. 8.5). At the front edge of the marine air is the sea breeze front.

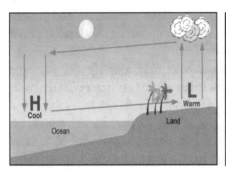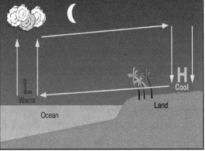

Fig. 8.4 Land and sea breeze circulation shows surface onshore flow during the day and offshore flow at night. Courtesy of Michael Pidwirny

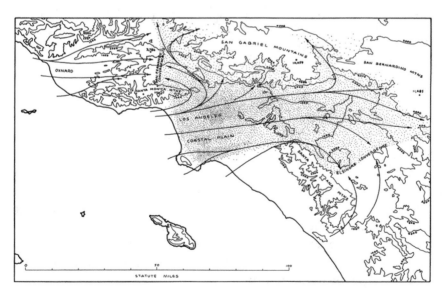

Fig. 8.5 Sea breeze flow into the San Fernando Valley. Note the San Fernando convergence zone, upper left, and the Elsinore convergence zone, lower right (Edinger and Helvey 1961)

The sea breeze also is responsible for air pollution patterns as daytime sea breezes push coastal city pollutants inland, while nighttime land breezes push pollutants back out toward the coast. With weaker winds, the same air pollution can remain over coastal valleys for days.

In the Los Angeles area, however, the Basin is almost completely enclosed by mountains on the north and east. Coupled with this is a dominant sea breeze pushing pollutants inland. The vertical temperature structure (inversion) tends to prevent vertical mixing of the air through more than a shallow layer (1000–2000 feet deep). With the concentrated population and industry, pollution products tend to accumulate and remain within this land-sea breeze circulation (See air pollution chapter for details).

Irregular or rough terrain in a coastal area may amplify the sea breeze front and cause convergence lines of sea breezes coming from different areas. Southern California and parts of the Hawaiian Islands are favorable for sea breeze soaring because orographic lift is added to the frontal convection. Sea breezes occasionally may even extend to the leeward sides of hills and mountains unless they are too high and long without breaks. Mostly, the sea breeze front converges on the windward slopes, and upslope winds add to the convection. Where terrain is fairly flat, sea breezes may penetrate inland for surprising distances but with weaker lift along the sea breeze front. Sea breezes reaching speeds of 15–25 knots (17–29 mph) are not uncommon.

In northern and central California, the onshore flow continues all day long although weakening at night. Further south, coastal areas experience a true land-sea breeze circulation.

8.7   Los Angeles "Smoke Front"

The ocean breeze often makes its way from the Los Angeles Basin into the Mohave Desert. Because it is often accompanied by much pollution, it has been called the "smoke front" or "smog front." This front leads to very interesting areas of convergence that can be seen in Fig. 8.5. Where the winds converge, strong vertical currents occur. The San Fernando and Elsinore convergence zones may result in lift that allows sailplanes to typically reach altitudes of between 6000 and 12,000 ft. Closer to the coast, there is little or no lift in the polluted ocean breeze air.

Prevailing wind influences the strength and direction of the sea breeze. A strong offshore (Santa Ana) could prevent the sea breeze from reaching the shore. On the other hand, a weak pressure gradient leads to onshore breezes that are perpendicular to the coastline. The urban heat island may enhance sea breezes when cities are along the coast, such as throughout southern and central California.

Forecasting the sea breeze consists of at least three major components (NWS): (1) a simple yes/no (on whether or not it will occur), (2) the wind speed and direction, and (3) the distance of inland penetration during the day.

8.8   Mountain Valley Winds

Unequal heating between mountain slopes facing the sun and shaded valleys can produce diurnal wind patterns in steep terrain. Especially in warmer months and longer daylight hours, an upslope flow (valley breeze) occurs along mountain slopes as the heated slope produces lower pressure and rising motions of less dense air. Rising air may even create clouds over the peaks if enough moisture is present and conditions are not too stable. At night, colder air near mountain tops flows by gravity downhill toward the valley (mountain breeze). During hot weather, people living along foothills enjoy this cooling mountain breeze (Fig. 8.6).

When coastal areas have complex terrain or mountains, the mountain valley winds interact with sea and land breezes. During warm summer days, sea breeze flows onshore and upslope. Then after sunset, cool downdrafts from the hills and mountains flows back toward the coast in a land breeze.

Fig. 8.6 Mountain and valley breezes. The first figure shows cool downslope mountain breeze. Second figure shows warm upslope valley breeze (Federal Aviation Administration, 2016)

8.9   Catalina Eddy

An eddy is a circular movement of air or water separated from the main current. The Catalina Eddy is so-called as it is centered near the Catalina Island off the coast of southern California.

When northwest wind along the coast interacts with the complex coastal terrain near Point Conception, it forms a counterclockwise low-pressure circulation. As northward moving air near San Diego meets the southward moving air near Santa Barbara a large eddy (the Catalina Eddy) may form (Fig. 8.7). This eddy may be as much as 200 miles in diameter. Wind speeds where the land juts outwards toward the west can be twice as great as winds at nearby points. The Catalina Eddy, a recurring counterclockwise circulation in the ocean off Los Angeles, is a challenge to local weather forecasters. This oceanic circulation upwells colder waters from lower oceanic depths producing fog and low clouds over the ocean. During an otherwise typical sunny June day, temperatures can drop twenty degrees when the Catalina Eddy is strong and brings in low clouds into the Los Angeles Basin. Beach-goers need sweaters and coffee not sunscreen. Eddy's smallness and usually shallow vertical extent make it difficult to monitor and predict. The Eddy can also transport Los Angeles smog northward into Santa Barbara's cleaner air.

Fig. 8.7 Catalina Eddy resulting from strong northwest winds brings low clouds and cooler temperatures to the southern California coastline (Alchetron.com)

8.10 Haboobs

Haboobs are common in the Arabian Peninsula and other parts of the Middle East In fact, the word haboob comes from the Arabic and means "violent wind." These winds are usually accompanied by extremely large walls of dust, hampering visibility. They also occur, however, in the American southwest including parts of southern California. Also called black blizzards, haboobs are produced by downdrafts flowing out of thunderstorms. When thunderstorms drop rain into dry air, the water evaporates and cools the surrounding air. The cool, dense air descends and spreads out at the ground picking up sand, dust, and other debris. Usually occurring in summer, hot air cools quickly. Dust rises thousands of feet into the air while moving like a gust front at near hurricane wind speeds. The visibility drops to a few feet. The sand and dust can be destructive at these speeds, pitting windshields, and sandblasting objects in its path. December 19–21, 1977, a strong haboob struck near Arvin, California at the south end of the San Joaquin Valley in California with wind gusts estimated at 192 mph. Known as the **Southern San Joaquin Valley Dust Storm,** dust from the haboob dimmed the sun as far north as Reno, Nevada. Sand and dirt accumulated to great amounts on highways and fields. Cars were buried and windows blown out. In parts of the valley, 70% of the homes were damaged. The wall of dust that results was estimated to be 5000 feet high. The event killed three people, caused nearly $120 million in 2002 dollars, not counting agriculture, with the National Weather Service ranking it as the ninth top weather event in California of the twentieth century. An outbreak of Valley fever (caused by a fungus that lives in the soil) followed as spores blew into Sacramento and Redding. A great ape at the San Francisco Zoo died from the fever.

A convergence of events led to the great dust storm. There had been a drought in the region for several years, which caused the ground to be dry with loose soil. Cotton had recently been plowed under (end of the season) but the winter crop had not taken root yet. The high winds were also caused by a strong pressure gradient, with cold, high pressure over the Great Basin and an advancing deep low off the northwest coast of California.

On November 29, 1991, a massive car and truck collision near Coalinga in the San Joaquin Valley killed 17 people. More than 100 vehicles were involved in the accident on Interstate 5, which was caused by a dust storm. Winds gusted to 50 mph, propelled sand and dust from nearby fields that were into their fourth year of drought. Visibility dropped to zero in the worst highway pileup in state history. The same stretch of highway was the scene of a similar, but smaller, incident in December 1978 when seven people died and 47 were injured in a large chain collision.

Land use can really affect dust storm frequency. Great examples of this are Owens Lake and the Salton Sea in California. Irrigation can affect the size of lakes in different ways. If the net effect is drying up the lake, then the exposed soil, usually unvegetated, contains fine material (as well as toxic fertilizer run-off, in the case of the Salton Sea), and can produce substantial amounts of dust. Owens Lake was the largest dust source in the United States at one point.

The southeastern California desert also experiences dust storms and sandstorms. The Palm Springs area was hit with a haboob Aug. 22, 2014, with 40 mph winds from thunderstorms coming from the North American monsoon felling trees and reducing visibility. Strong pressure gradients in spring between high pressure along the coast and low pressure over the hot desert are great for turning windmills north of Palm Springs, but may lead to damage and disruption to the resort area such as March 28, 2016, when massive car pileup injured 28 people. There have been as many as 34 dust storms in the Coachella Valley since 1993 causing poor visibility and unhealthy air quality (Fig. 8.8).

Fig. 8.8 Haboobs over southeast California and in the Central Valley can reduce visibility to near zero (NOAA, 2012)

8.11 Rotor Winds

Besides Santa Anas and other downslope winds, mountains can cause local mountain valley winds and rotor winds. The rotor wind is a form of lee wave such as a mountain wave that forms a turbulent vortex downwind of a mountain range. In a rotor, the wind at the ground blows toward the mountain. One of the most spectacular lee waves is the **Sierra Wave**, which occurs when westerly winds flow over the Sierra Nevada Range in California. It is best developed when the polar-front jet stream blows across the range. In it, gliders have soared to elevations of more than 14,000 m (near 46,000 ft). But it can be hazardous to unwary pilots. These turbulent lee wave winds are especially dangerous in aviation. Rotor winds may be identified by mountain wave clouds, such as lenticulars (Fig. 8.9), which form as mountain waves lift and cool air to condensation in the updraft portion of waves.

Also east of the Sierras, downslope (katabatic) windstorms associated with strong pressure gradients of passing storms can reach 100 mph in winter and spring when the jet streams cross northern California.

Fig. 8.9 Lenticular clouds form saucer-like stacks downwind of mountains, like the Sierras shown here (Photo by Bill Reid)

8.12 Sonora Wind

Another hot, dry wind is the **Sonora** that flows from Sonora, Mexico, into Baja California and southern California in late spring.

8.13 Tropical Storms

Tropical storms can produce damaging winds and remain a threat mainly in southern California. This topic is covered in Chap. 9. Tropical Pacific storms off the west coast of Mexico occur frequently from May through November, though few reach near the California border. Hurricane Nora reached the eastern edge of southern California near Yuma, Arizona on Sept. 25, 1997. Although Nora's winds did not significantly impact any area of southern California, the rainfall certainly did.

More recently, on September 8th and 9th 2022, Tropical Storm Kay affected southern California. Although Kay passed to the southwest of the California coast, peak winds of 109 mph from the east were recorded at Cuyamaca Peak east of San Diego. Elsewhere in the San Diego area, wind gust as high as 60 mph were common.

Bibliography

Alchetron. (2023). *Catalina eddy—Alchetron, the free social encyclopedia.* https://alchetron.com/Catalina-eddy

Blier, W. (1998). The sundowner winds of Santa Barbara. *Weather & Forecasting, 13,* 702–716.

Chandler, R. (1946). *Red wind: A collection of short stories.* World Publishing Company.

Dorman, C. E., Klimczak, E., & Nuss, W. (2002). *The Catalina Eddy and topographic forcing over the Southern California bight.* https://www.researchgate.net/profile/Clive_Dorman/publication/320434560_The_Catalina_Eddy_and_Topographic_Forcing_over_the_Southern_California_Bight/links/59e550660f7e9b0e1aa88d5f/The-Catalina-Eddy-and-Topographic-Forcing-over-the-Southern-California-Bight.pdf

Edinger, J. G., & Helvey, R. A. (1961) The San Fernando convergence zone. *Bulletin of the American Meteorological Society, 42,* 626–635.

Federal Aviation Agency. (1975). *Aviation weather-sea breeze soaring.* Superintendent of Documents, US GPO, Washington, DC. https://www.aviationweather.ws/097_Sea_Breeze_Soaring.php

Federal Aviation Agency. (2016). *AC 006B aviation weather*. https://www.faa.gov/ documentlibrary/media/advisory_circular/ac_00-6b.pdf

Finley, J., & Raphael, M. (2007). The relationship between El Niño and the duration and frequency of the Santa Ana winds of southern California. *The Professional Geographer, 59*(2), 184–192.

Garza, A. L. (1999). *1985–1998 North Pacific tropical cyclones impacting the southwestern United States and northern Mexico: An updated climatology*. NOAA Tech Memo NWS WR-258.

Goodridge, J. D., Rhodes, H., & Bingham, E. G. (1979). *Windstorms in California*. Department of Water Resources, State of California.

Great Duststorm of 1977. (2022, July 5). Wikipedia. https://en.wikipedia.org/wiki/ Great_Bakersfield_Dust_Storm_of_1977

Guzman-Morales, J., Gershunov, A., Theiss, J., Li, H. Q. & Cayan, D. (2016). Santa Ana winds of Southern California: Their climatology, extremes, and behavior spanning six and a half decades. *Geophysical Research Letters, 43*(6), 2827–2834. https://doi.org/10.1002/2016GL067887

Kaplan, C., & Thompson, R. (2009). *The Palmdale wave: An example of mountain wave activity on the lee side of the San Gabriel Mountains*. Western Regional Technical Attach Lite No. 09-05. https://www.weather.gov/media/wrh/online_public ations/talite/talite0905.pdf

Mensing, S. A., Michaelsen, J., & Byrne, R. (1999). A 560-year record of Santa Ana Fires reconstructed from charcoal deposited in the Santa Barbara Basin, California. *Quarterly Research, 51*, 295–305.

Miller, S. T. K, Keim, B. D., Talbot, R. W., & Mao, H. (2003). Sea breeze: Structure, forecasting and impacts. *Reviews of Geophysics, 41*(3). http://doi.org/10. 1029/2003RG000124

National Weather Service. (2023). *Katabatic winds*. https://katabaticwinds.weebly. com/types.html

NOAA. (2012). *Haboobs: Phenomena with the unusual name is no joke*. https://www. noaa.gov/stories/haboobs-phenomena-with-unusual-name-is-no-joke

NOAA. (2023). *Daily weather maps*. https://www.wpc.ncep.noaa.gov/dailywxmap/

Null, J. (2015). Weather and wildland fires: Fighting fires in the age of droughts and urban sprawl. *Weatherwise, 68*(4), 28–33.

Parzybok, T. W. (2005). *Weather extremes in the west*. Mountain Press.

Pidwirny, M. (2018). *Fundamentals of physical geography* (2nd ed.). http://www.phy sicalgeography.net/fundamentals/7o.html

Rauber, R. M., Walsh, J. E. & Charlevoix, D. J. (2019). *Severe and hazardous weather: An introduction to high impact meteorology* (5th ed.).

Rolinsky, T., Scott, B. C., & Zhuang, W. (2019). Santa Ana winds: A descriptive climatology. *Weather & Forecasting, 34*, 257–275. https://doi.org/10.1175/WAF-D-18-0160.1

Ryan, G. (1996). *Downslope winds of Santa Barbara County, California*. Western Region Technical Memorandum NWS WR-240, National Weather Service, Oxnard, California. https://repository.library.noaa.gov/view/noaa/14746

Sukup, S. (2014). *Damaging downslope wind events in the San Gabriel Valley of Southern California.* Western Region Technical Attachment No. 16-02. https://www.weather.gov/media/wrh/online_publications/TAs/TA1602.pdf

Tompkins, W. A. (1976). *It happened in Old Santa Barbara.* Santa Barbara National Bank.

Vasquez, T. (2008). The ill wind that blows: Southern California's Santa Ana phenomena. *Weatherwise, 61*, 34–39.

Wakimoto, R. M. (1987). The Catalina Eddy and its effect on pollution over Southern California. *Monthly Weather Review, 115*, 837–855.

Westerling, A. L., Cayan, D. R., Brown, T. J., Hall, B. L., & Laurence, G. R. (2004). Climate, Santa Ana winds and autumn wildfires in Southern California. *Eos, 85*(289), 296.

9

Cyclones: Mid-latitude and Tropical

26 Nov 2019 20:10Z NESDIS/STAR GOES-West GEOCOLOR

Fig. 9.1 A Pacific cyclone takes aim on California bringing strong winds and heavy precipitation. In this infrared image, you can see the counterclockwise rotation of winds and clouds, a distinct cold front slicing through the state, and coldest (more precipitation) clouds behind the cold front (Image from November 26, 2019-NOAA)

© The Author(s), under exclusive license to Springer Nature
Switzerland AG 2023
S. LaDochy and M. Witiw, *Fire and Rain*,
https://doi.org/10.1007/978-3-031-32273-0_9

"The general setup seemed to indicate that in the next twenty-four hours the new storm would move rapidly eastward. As it moved, it could grow in both area and intensity; its winds becoming stronger, its rains heavier." from **Storm** (1941), George Stewart's fictional account where he tracks a cyclone that makes its way from the Pacific Ocean through California with strong winds, heavy rain, and deep mountain snow (Fig. 9.1).

9.1   Surface Weather Patterns

The Mediterranean climate of coastal California is a result of seasonal shifts in pressure and wind patterns. In summer, a dominant North Pacific High-pressure system moves north, blocking storms from following westerlies into the state (Fig. 9.2). In winter, the Pacific High shrinks and moves south, while a vigorous Aleutian Low strengthens in the Gulf of Alaska. Winter precipitation in California mostly originates with the Aleutian Low.

9.2   Mid-latitude or Frontal Cyclones

A front is a boundary between cold and warm air. It forms in the mid-latitudes, the region where you get the strongest contrast between cold air to the north and warm air to the south. Disturbances may form on that boundary and sometimes grow and develop into frontal cyclones. These storms have a life cycle of several days. Pressures fall in the new cyclone and a center of low pressure strengthens with winds moving counterclockwise in the Northern Hemisphere. In general, cold air moves south, while warm air moves north, resulting in two fronts. A cold front, indicated on weather charts by triangles, is when the cold air is advancing replacing warm air. A warm front, indicated by the half-circles on weather charts, is when the warm air is advancing replacing colder air (Fig. 9.4).

These mid-latitude or frontal cyclones are large traveling atmospheric areas of low pressure that are up to about 1200 miles in diameter. An intense mid-latitude cyclone may have a surface pressure as low as 980 mb (millibars), compared to an average sea-level pressure of 1013 mb. Normally, individual frontal cyclones exist for about three to 10 days. They form in the belt of westerly winds and therefore generally move west to east in both the northern and southern hemispheres. Frontal cyclones are the dominant weather event of Earth's mid-latitudes.

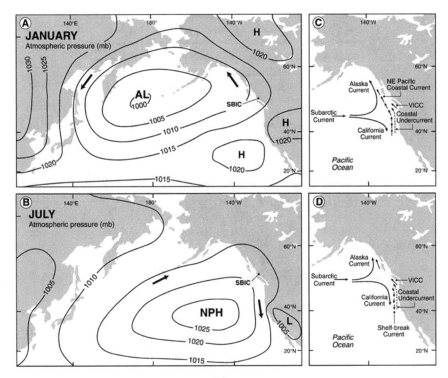

Fig. 9.2 Typical location of the pressure systems in the North Pacific Ocean in winter and summer. "AL" refers to the low-pressure "Aleutian Low" and "NPH" refers to the high-pressure "North Pacific High" system. The clockwise winds around the NPH drive the California Current southward, with a stronger current in spring and summer (NOAA, Office of Response and Restoration)

9.2.1 Life Cycle of Mid-latitude Cyclones

A cyclone typically begins as a weak disturbance somewhere along a frontal boundary where initially the boundary between the cold and warm air is relatively stationary. As the disturbance strengthens, pressure falls, and warm air is lifted over the cold air. As a result, both a warm and cold front form as the counterclockwise circulation around the new low-pressure system pushes warm air ahead of the low and pulls cold air to its rear. As the cyclone matures and pressure falls, the faster moving cold front eventually overtakes the warm front. The result is an occluded front. Along the occluded front, there is no longer any warm air at Earth's surface. It has all been lifted aloft (Figs. 9.3, 9.4 and 9.5).

Once the front is occluded, the associated low-pressure area is no longer on the frontal boundary and pressures begin to rise as the low weakens and eventually dissipates.

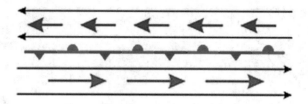

Fig. 9.3 Stationary front separating cold from warm air with the cold air to the north in the Northern Hemisphere (NOAA Jetstream)

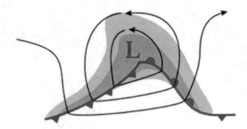

Fig. 9.4 Mid-latitude cyclone with warm and cold front (NOAA Jetstream)

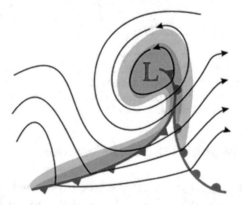

Fig. 9.5 Occluded front shown to the north of where the cold front and warm front meet (NOAA Jetstream)

Cyclones originating over the Pacific can strike anywhere along the Pacific coastline but are most common from northern California northward. In winter, storms form in the Gulf of Alaska from the semi-permanent Aleutian Low (Fig. 9.2). These lows usually move to the southeast bringing moisture to California. The north–south oriented mountains along the California coast and especially along the Sierras lead to orographic lifting and increased precipitation compared to the coast. Cyclones weaken as they move over the mountains.

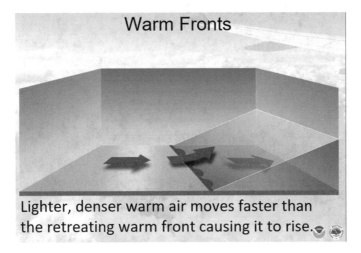

Warm Fronts

Lighter, denser warm air moves faster than the retreating warm front causing it to rise.

Fig. 9.6 Warm front showing warm air overrunning colder air (NWS, NOAA)

9.2.2 Warm Frontal Weather

Ahead of a warm front, humid air is transported upward over a distance that may exceed 600 miles. As a warm front approaches, upper clouds (cirrus and cirrostratus) begin to move in. As the front gets closer, clouds will lower and thicken. The high clouds will eventually give way to mid-level altostratus clouds and then, lastly to precipitating nimbostratus clouds. Warm fronts typically move slower than cold fronts and precipitation tends to be long-lasting and over a wide area. Temperatures will slowly rise ahead of a warm front. After its passage winds will shift (frequently from the south to south-west), humidity and temperatures will rise, and steady precipitation will end. Along the west coast as the low-level air is moderated by the ocean, we usually see relatively small rises in temperature at sea level (Fig. 9.6).

9.2.3 Cold Frontal Weather

Unlike a warm front, the weather that accompanies a cold front into California is quite different from what we see east of the Rockies in the United States. East of the Rockies, showers and thunderstorms usually precede a cold frontal passage with clearing to its rear. However, in California, for the large majority of cold fronts that come in from the Pacific Ocean, continuous light to moderate rain precedes the front. After frontal passage cold air moves in at higher altitudes, but the influence of moderate ocean temperature usually prevents any significant cooling near sea level. Along coastal California, often

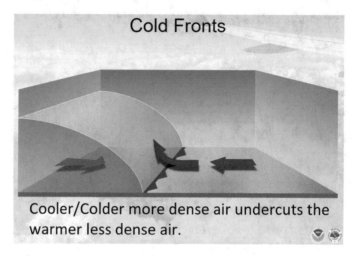

Fig. 9.7 Cold front showing cold air moving in from the left forcing warm air ahead of it upwards (NWS, NOAA)

there will be unstable, cold air aloft behind the cold front resulting in showers and possibly severe weather after the cold front passes. This is unlike cold front weather over central and eastern United States when clearing typically occurs behind a cold front. The cold air aloft results in lowering snow levels at higher elevations. The snow is frequently heavy in the showers that occur after passage of a maritime cold front. An exception is the rare cold front that brings continental air from the north. These continental fronts bring significant cooling and generally clear to partly cloudy skies after passage. They are quite rare and may not occur in many winters (Fig. 9.7).

9.2.4 Occluded Front

A cold front will move faster than a warm front, eventually catching up with it. When this happens, the front is said to be occluded. In areas where an occlusion exists, all the warm air is forced aloft. This additional lift increases precipitation in its vicinity. Precipitation is similar to that occurring with both warm and cold fronts, that is, steady light to moderate precipitation with an occasional embedded heavy shower. An occluded front is represented by a combination of warm and cold front symbols on weather maps.

Figure 9.8 shows both a warm and cold occlusion. In a cold occlusion, the air behind the occlusion is colder than that ahead of it. The converse is true in a warm occlusion. Occlusions can also be neutral with similar air temperatures on both sides.

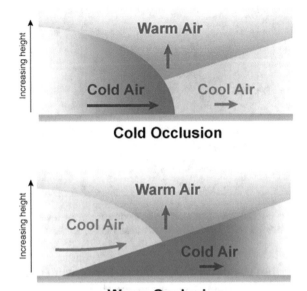

Fig. 9.8 Occluded fronts (NOAA Jetstream)

9.3 Cutoff Lows

Once a frontal system becomes occluded, the low-pressure area is no longer present on the frontal boundary. It is essentially detached from the front as a cutoff low-pressure area. At this point, the low (cyclone) weakens and pressure rises. Occasionally cutoff lows at Earth's surface will develop in response to low pressure high in the troposphere. Precipitation associated with cutoff lows tends to be showery rather than steady. Cutoff lows along the coast of California may lead to abundant rainfall due to its ocean source of moisture (Fig. 9.9).

9.4 Jet Stream

The polar jet stream is a ribbon of fast-moving wind in the upper troposphere at about seven to nine miles above sea level. The polar jet stream marks the boundary between cold air to the north and warmer air to the south. Since the polar front also occurs at this boundary, they are found in the same general location with the jet stream a bit poleward of the polar front. The seasonal shift in storm tracks along the west coast mirrors the southward shift of the

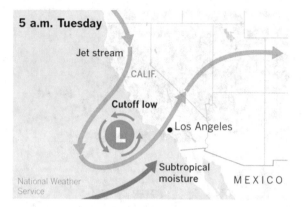

Fig. 9.9 A winter cutoff low spins inside a deep upper-level trough, while an atmospheric river brings rain from the southwest (NWS, NOAA)

polar jet stream from the northwest in fall and early winter to southern California by February. In spring, the jet stream shifts back north as do storm tracks. As storm tracks follow the jet stream, it is not surprising that precipitation peaks as the jet stream moves over a region. Hence northern California peaks earlier, December and January, while further south the rainfall peak is usually in February.

The polar jet stream often develops waves such that troughs (dips southward) and ridges (northward bulges) can occur. Upper air troughs are associated with surface low pressures, while ridges are above surface high pressures (Fig. 9.10). When a ridge remains over the west coast, it blocks storms from bringing precipitation to California. If this persists, droughts may occur.

During El Niño years, the polar jet stream is displaced farther south than average in winter, thus directing storm tracks across the southwestern United States and producing wetter than average conditions there and in Baja California.

9.5   Memorable Mid-latitude Cyclones

Cyclones get their strength from the pressure differences between the cyclone low-pressure center and that of its surrounding, or its pressure gradient. The greater the pressure gradient, the greater are the cyclonic winds associated with the storm. Occasionally, storms may intensify into hurricane strength winds. Below are some historically important storms.

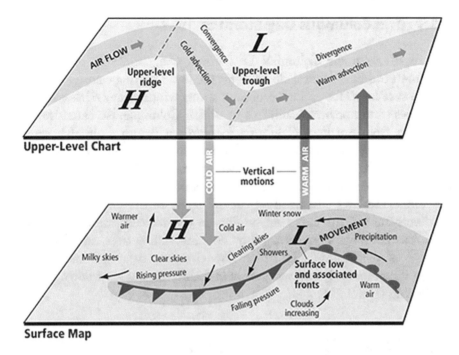

Fig. 9.10 Upper air divergence leads to surface convergence and development of frontal cyclones (Courtesy *Ocean Navigator*)

9.5.1 Thanksgiving Day Storm of 2019

The lowest barometric pressure ever recorded in the state of California was at Crescent City at 973.4 mb on a Thanksgiving Day (Nov. 26, 2019) breaking a record from Jan. 20, 2010. The storm resulted in heavy snow, hail, rain, and winds to northern California shutting down Interstate 80 near Tahoe. Winds hit 69 mph at the Crescent City Airport, with gusts up to 80 mph nearby. The storm originated out of the Gulf of Alaska as a "bomb," a cyclone that rapidly intensified with falling pressure within 24 h.

9.5.2 January 20, 2010

Further south, the Jan. 20, 2010 storm broke all-time low pressure records at San Diego Lindbergh Field (29.28 in. or 982 mb), while Los Angeles International Airport broke their record at 980 mb or 29.08 in. Hg using a standard mercury in glass barometer.

9.5.3   The Columbus Day Storm of 1962

The Big Blow or the Columbus Day Storm, Oct. 12, 1962, probably the most powerful non-tropical storm to hit the lower 48 in a century with 955 mb pressure low. This cyclone produced strong winds, heavy rains and snow, mudslides from northern California to British Columbia. An extensive area, stretching from northern California to southern British Columbia experienced hurricane-force winds, massive treefalls, and power outages (Fig. 9.10). Sustained winds reaching 60–70 mph with gusts over 120 mph. Flooding and landslides in northern California were also a problem. In the San Francisco Bay area, landslides and flooding cut off access to the Orinda side of the Caldecott Tunnel, the Cull Canyon dam in Castro Valley nearly burst, and numerous buildings were damaged by flooding, falling trees, and landslides. Game 6 of the World Series in San Francisco was delayed because of the storms. The Nimitz Freeway in northern California was shut down because of flooding (Fig. 9.11).

9.6   Tropical Cyclones (Also Known as Chubascos)

Due to hurricanes needing close to an 80 °F sea surface temperature to form, California's cool ocean current is not suitable for tropical cyclones. However, it has happened. In September 1939, a tropical storm made landfall near Long Beach. The storm still had 50-mph winds at shoreline, with copious rainfall in the area, with much more in the surrounding mountains. Forty-five people were killed ashore and many more at sea. An earlier report of a hurricane reaching San Diego took place in 1858. These infrequent storms from the south were also referred to as Sonoras or Chubascos.

The cold current off southern California typically stays in the 60s (°F), cold enough to quickly weaken any hurricane or tropical storm that moves up from the waters west of Baja California. The Gulf of California is much warmer and provides the energy to sustain tropical cyclones when they reach this far north. The Gulf of California water can reach 90 °F in summer. However, most tropical storms reaching the Gulf tend to move northeast toward Arizona rather than populated coastal southern California. Additionally, the Gulf is rather narrow (at its widest, it is about 120 miles wide) and because of that it is very rare for a tropical cyclone to remain over it for any length of time. But there are exceptions.

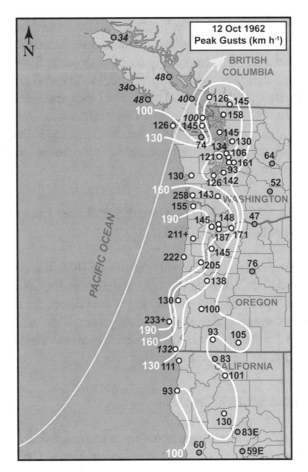

Fig. 9.11 Peak wind gusts (km/h) from The Columbus Day Storm, Oct. 12, 1962. The yellow arrow shows the path of the storm (Office of the Washington State climatologist)

During El Niño years, the frequency of eastern Pacific tropical cyclones tends to increase. The number of tropical cyclones that affect the southwest United States is about one per year increasing to two to three times per year during El Niños. Some of these tropical cyclones affected southern California, particularly desert and mountainous areas. When these storms affect southern California, winds are seldom a problem, but heavy rainfall may occur.

Since 1900, four of those storms brought gale-force winds to California: an unnamed California tropical storm in 1939, Tropical Storm (TS) Kathleen in 1976, TS Nora in 1997, and TS Kay in 2022 While dissipating tropical cyclones reaching California will not have the wind strength of the more powerful Santa Ana winds, they may result in excessive rainfall, with local flooding as upper air winds may bring clouds and moisture from the cyclone over southern California to produce showers and thunderstorms. During late summer and early fall, tropical cyclones may reach further northward, recurving to the northeast when upper troughs extend further south into lower latitudes. In September1997, Hurricanes Linda and Nora both brought heavy surf and heavy rainfall to southern California coasts. Details on some of the more notable storms follow. As weather data coverage increased and satellite imagery became available, more complete information on these tropical cyclones became possible.

9.6.1 The San Diego Hurricane of 1858 and Long Beach Tropical Storm of 1939

Details are lacking on the San Diego Hurricane of 1858, but no hurricane has made landfall on the California coast since that year. Reports suggest winds of 75 mph with uprooted trees and houses blown down. Another direct hit, or landfall, by a tropical cyclone was the Long Beach Tropical Storm of 1939 causing torrential rains with Los Angeles receiving over five inches. Numerous deaths occurred at sea and due to flooding (Fig. 9.12).

9.6.2 Tropical Storm Kathleen

TS Kathleen, Sept. 1976, was one of only two tropical storms, besides Nora, to retain its power as far as the southwest United States. Downgraded from hurricane to tropical storm, Kathleen crossed the border near Calexico, California, moving into Death Valley and then western Nevada on Sept. 10. Heavy rains and winds up to 76 mph damaged Yuma with millions in damages. Up to 10 inches of rain flooded the southern California deserts. Major damage took place in Ocotillo, California with extensive flooding and sediments.

9.6.3 Hurricane Linda

Although passing well to the southwest of California, Hurricane Linda did have a major impact on southern California. Hurricane Linda was the second-strongest eastern Pacific hurricane on record at the time. Though it remained far from the state, the hurricane produced light to moderate rainfall across the region, causing mudslides and flooding causing millions in damages. Huge waves also pounded the southern California beaches.

Fig. 9.12 Tropical cyclones that have affected southern California and their storm paths (NOAA)

9.6.4 Hurricane Nora

Hurricane Nora became a tropical storm when reaching southern California from September 22 to September 26, 1997, rainfall totals ranged from nearly two inches in the desert areas near Palm Springs and Thermal to 5.50 inches at Mount San Jacinto. Several desert roads were flooded out. Waves, up to 20 feet, battered Orange County beaches. Damages from the storm reached hundreds of millions of dollars, mostly to agriculture (Fig. 9.13).

9.6.5 Tropical Storm Kay

More recently, on September 8–9, 2022, TS Kay affected southern California. Unusual September rainfall affected much of the area, with San Diego's Lindbergh Field reporting over 0.60 in. of rain and Palm Springs 0.50 in. on September 9. A few showers continued over mountainous area occurred through September 12. Winds of 30–50 mph affected the San Diego area. More on the strong winds associated with Kay are addressed in Chap. 8 (Fig. 9.14).

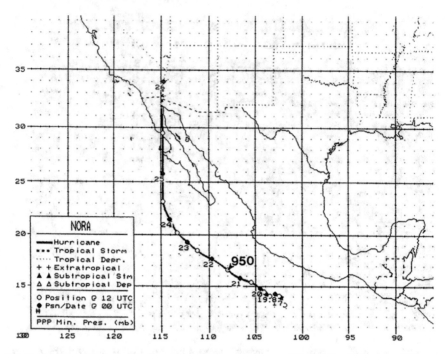

Fig. 9.13 Plot from the National Hurricane Center archives shows TS Nora crossing into the southwest United States (National Hurricane Center)

Fig. 9.14 TS Kay at its closest approach to southern California. ♦ is the symbol for a tropical storm (chart from NOAA, 2023 Daily weather maps)

Bibliography

Barton, N. P., & Ellis, A. W. (2009). Variability in wintertime position and strength of the North Pacific jet stream as represented by re-analysis data. *International Journal of Climatology, 29*, 851–862.

Belmechen, S. (2017). Northern hemisphere jet stream position indices as diagnostic tools for climate and ecosystem dynamics. *Earth Interactions.* https://doi.org/10.1175/EI-D-16-0023.1

Garza, A. L. (1999). *1985–1998 North Pacific tropical cyclones impacting the southwestern United States and northern Mexico: An updated climatology.* https://www.weather.gov/media/wrh/online_publications/TMs/TM-258.pdf

Manney, G. L., & Heglin, M. I. (2018). Seasonal and regional variations of long term changes in upper tropospheric jets from reanalyses. *Journal of Climate, 31*, 423–448.

National Hurricane Center. (1997). *Preliminary report Hurricane Nora.* https://www.nhc.noaa.gov/data/tcr/EP161997_Nora.pdf

National Weather Service. (2023). *Weather history.* https://www.weather.gov/sgx/weather_history

NOAA. (2022). *GOES image viewer.* https://www.star.nesdis.noaa.gov/goes/conus_band.php?sat=G17&band=GEOCOLOR&length=12

NOAA. (2022). *Office of response and restoration.* https://response.restoration.noaa.gov/

NOAA. (2023). *Daily weather maps.* https://www.wpc.ncep.noaa.gov/dailywxmap/

NOAA National Weather Service. (2022, December 9). *Jetstream—An online school for weather.* https://www.weather.gov/jetstream/

Oakland Wiki. (2022, December 9). *Storms of 1962.* https://localwiki.org/oakland/Storms_of_1962

Ocean Navigator. (2017). *The cut-off low.* https://oceannavigator.com/the-cut-off-low/

Office of the Washington State Climatologist. (2015). *The Columbus day storm: A perspective by Wolf Read.* https://www.climate.washington.edu/stormking/October1962.html

Raphael, M. N., & Cheung, I. K. (1998). North Pacific midlatitude cyclone characteristics and their effect upon winter precipitation during selected El Nino/southern oscillation events. *Geophysical Research Letter, 25*, 527–530.

Raphael, M. N., & Mills, G. M. (1996). The role of mid-latitude Pacific cyclones in the winter precipitation climate of California. *Professional Geographer, 8*, 251–262.

Steenburgh, W. J., & Mass, C. F. (1996). Interaction of an intense extratropical cyclone with coastal orography. *Monthly Weather Review, 12*, 1329–1352.

Stewart, G. (1941). *Storm.* Random House.

Strong, C., & Davis, R. E. (2008). Variability in the position and strength of the winter jet stream cores related to Northern Hemisphere teleconnections. *Journal of Climate, 21*, 584–592.

10

Thunderstorms and Thunderstorm-Associated Severe Weather

Fig. 10.1 Lightning on Santa Monica Bay as seen from Malibu (Courtesy of Bill Reid)

Thunder only happens when it rains…
—Dreams, Fleetwood Mac

10.1 Thunderstorms

A thunderstorm is produced by a cumulonimbus cloud. Thunder and lightning define a thunderstorm. Thunderstorms contain many hazards. Lightning is present in all thunderstorms, and many people are injured each year, with 80 being killed. Those who have been injured often experience chronic medical conditions. In addition to lightning, thunderstorms often contain other dangers to include microbursts, tornadoes, large hail, and sometimes result in flashfloods. Thunderstorms are extremely hazardous to aviation often producing severe icing and turbulence and must be avoided (Fig. 10.1).

Rain, often heavy, also usually accompanies a thunderstorm, although in some very dry areas, the rain may evaporate before reaching the ground. These types of "dry" thunderstorms are prevalent in parts of the western United States. Although rain may not reach the ground, lightning often does and sometimes results in wildfires.

Thunderstorms develop from warm, moist, unstable atmospheric conditions. Within this unstable atmosphere, we can trace parcels or bubbles of air that may be of different sizes—typically up to about five miles across. With sufficient heating, a moist air bubble rises. As it rises, the air cools. If cooling causes the parcel's rising air temperature to reach the dew point, condensation occurs, and a cloud forms. In unstable air, saturated air parcels can rise as long as they remain warmer than the surrounding air. Under extremely unstable conditions, clouds can grow into large cumulonimbus (thunderhead) clouds 10s of thousands of feet high. In the mature stage of development, thunderstorm clouds produce precipitation along with strong updrafts and downdrafts. Positive and negative charges move to opposite ends of these clouds producing a discharge of static electricity, lightning. Lightning defines a thunderstorm.

Thunderstorms may occur in California at any time of the year. Along the coast, thunderstorms are confined to the cooler winter and spring months, while in the deserts and mountains, there tends to be cool season and summer thunderstorms, with the latter fed by Gulf of California moisture during the Southwest monsoon (also called the North American monsoon). Coastal southern California stations show a peak in thunderstorm days in February or March, while most desert locations peak in July and August. The frequency of thunderstorms is also much higher in the mountains and deserts than along the coast. There are fewer weather stations in the mountains, but they experience thunderstorms caused by both winter storms and summer convection. Over the interior mountain areas and deserts, storms are

Annual Mean Thunderstorm Days (1993-2018)

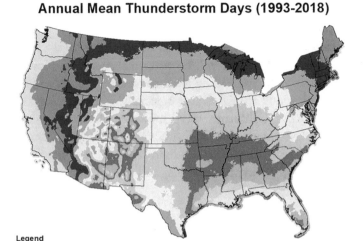

Fig. 10.2 This figure shows the relatively small number of thunderstorm days in California (NOAA Jetstream)

more intense, and they may become unusually severe on occasion at intermediate and high elevations of the Sierra Nevada. In these mountainous areas, thunderstorms average up to 27 days per year and frequently exceed that number. The frequency of southern California thunderstorms closely follows elevation, increasing from five days or less annually along the coast to the numbers seen over mountainous areas. Figure 10.2 puts California thunderstorm occurrence into perspective. It can be seen that for west coast states, including California, thunderstorms are relatively rare compared to the rest of the United States.

10.1.1 North American Monsoon

In southern California, the summer Pacific High blocks most storms from entering the coastal region. Thunderstorms, though infrequent occur mainly in winter storms along the coast peaking in February or March. Inland, the North American monsoon may bring thunderstorms, sometimes quite severe, to the mountains and deserts. Lightning may be accompanied by heavy downpours, damaging winds and the occasional tornado.

During the period from July through September, the North American monsoon is active. Tropical moisture is transported from the Gulf of California into the southern California desert and mountainous areas. This moist air sometimes raises the dew point to 70 °F. This moist, unstable air often results in thunderstorms.

The monsoon intensity varies from year to year. If there is a trough of cool air along the eastern Pacific, it will be weak. In 2020, the North American monsoon was nearly absent, leading to extreme drought and a record wildfire season. If coastal winds are light, and especially when tropical disturbances travel up the Gulf of California, more moisture travels into the deserts of Arizona and parts of southern California. Weather hazards associated with the summer monsoon storms include lightning and dry lightning wildfires, flash floods, damaging microburst winds, and the occasional tornado.

10.1.2 Topography

Topography plays a large role in thunderstorm generation and distribution throughout the state. Lifting of warm, moist air by mountains may trigger rapid cumulus cloud growth often developing in cumulonimbus clouds. As low-level air is funneled through complex terrain, convergence of this air will occur, resulting in lifting of heated, moist air resulting in summer inland thunderstorms.

On August 17, 1959, a heavy thunderstorm dropped 1.5 in. of rain in Needles and one inch at the Needles airport. One person died and three were missing in disastrous flooding. Bridges, highways, and railroads were washed out across a wide area. Cars were swept away. Waves up to 22 ft were observed coming down Sacramento Wash. This was considered the greatest runoff of record from any desert watershed in San Bernardino County history. Another moisture surge from the Gulf of California on August 29, 1993, showed that local convergence due to desert topography can lead to severe weather. A convergence zone occurring near Salton Sea produced an F1 tornado in the eastern Imperial Valley. Dew point temperatures can reach up to 70 °F in the southeastern deserts of California making the usual hot conditions steamy and uncomfortable.

August 14, 1990, a flash flood from thunderstorms washed out roads in the 29 Palms area of the southeastern desert. An hour earlier a funnel cloud was spotted near Victorville and a tornado touched down two miles east of Daggett, all in the San Bernardino County. Hail up to 1.75 in. in size caused extensive damage in the El Centro area of Imperial County.

10.1.3 Convergence Zones

Where low-level winds converge from different directions, the air will be forced upward. Varying terrain often is a cause of this convergence.

10.1.3.1 Noteworthy Convergence Zones

An interesting example a convergence zone is the phenomenon known as the Eastern Antelope Valley Eddy. A convergence zone sometimes forms between the westerly sea breeze winds that pass through the San Gabriel Mountains, and the southerly sea breeze that moves through the Cajon Pass. As these winds converge, the moist oceanic air is forced upward. When there is adequate instability, thunderstorms may form.

10.1.3.2 Island Effect Storms

Ivory Small in 1999 found coastal islands in the southern California region can create convergence zones downwind with northwest wind flow, sometimes leading to thunderstorms and severe weather. This phenomenon is known locally as the "Island Effect."

A significant Island Effect storm occurred on November 10, 2000, when strong precipitation bands developed by island-generated convergence. During this time, a tornado occurred in the city of Poway, about 20 miles northeast of downtown San Diego. The tornado formed on a cloud band that extended downwind from the islands (basically an "island effect" tornado).

10.2 Lightning

By definition, all thunderstorms must have lightning. Most lightning, however, do not reach the ground. In California, cloud-to-ground lightning can cause power outages, property damages, wildfires and personal injuries and fatalities. California ranks 29th in the United States for lightning deaths, averaging eight fatalities per year, based on records from 1990 to 2003. The ranking falls to 49th when considering deaths per million due to the state's high numbers of people. No fatalities occurred in the state in 2017 when only 16 deaths occurred from lightning nationwide, a record low. This follows a trend of decreasing lightning deaths since the 1940s when lightning was the leading cause of weather-related death. Injuries, though, much more frequent, have also decreased nationally to 120 reports in 2016. Injuries by lightning are thought to be greatly underreported. Some significant events follow.

In April 1926, lightning struck an oil storage facility near San Luis Obispo. A fire that lasted five days resulted in over 900 acres of land being devastated and over six million barrels of oil destroyed. Two people were killed, and damage was estimated at 15 million dollars.

Eight members of a girls' softball team and an adult coach were injured in a Tustin schoolyard April 23, 1988, when lightning triggered by a fast-moving storm that raked the Southland struck a tree under which they had sought refuge from the rain.

On a day that started out sunny, thunderstorms swept into southern California on July 27, 2014, causing 12 injuries and an apparent death along Venice Beach when lightning hit the water.

A three-day series of summer thunderstorms in early August 2017 initiated 2100 lightning strikes in the southern California region including setting off fires, widespread power outages and in other areas heavy rainfall. A map showing the annual average flash density in the southwest California area, excluding San Diego's forecast area, shows the highest rates in mountainous areas (Fig. 10.3). These are mainly associated with the summer monsoon.

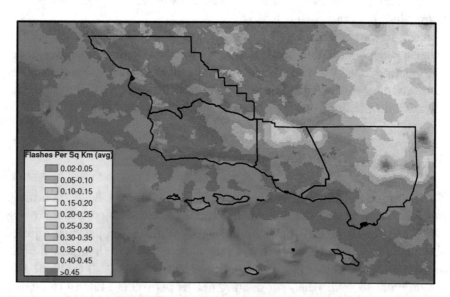

Fig. 10.3 Average annual lightning flash density in the Los Angeles area (Meier & Thompson, 2009, Western Regional Headquarters)

Lightning in the San Francisco Bay area caused outages to over 1000 customers of PG&E and brought down a 80-ft redwood into a house in Sausalito on Sept. 11, 2017. Over 1200 lightning strikes were seen over the Bay area that day, delaying the Giants baseball game at AT&T Park. Showing its attraction to metals like transmission lines, a West Covina lady was struck in her kitchen washing dishes when lightning traveled along the plumbing into her house.

Figure 10.3 shows the average annual number of lightning flashes per square kilometer. Flash density increases 20-fold from coastal to mountainous areas. Local maxima in flash density can be found over the Mt. Pinos area in Ventura County, and over the eastern San Gabriel Mountains and Antelope Valley in Los Angeles County (data from 1987 to 2000).

Lightning occurs most often in mountainous areas, like the Sierras, being a hazard to campers and hikers. Figure 10.4 shows August 15, 2020, thunderstorms causing 1089 lightning strikes. On this day the devastating Mendocino August Complex fires started in northern California. This was followed by the Santa Clara Unit (SCU) Lightning Complex fires on August 16, 2020, with 20 more separate fires. The SCU Fires were the fourth largest fires in California history (see Appendix C). Especially in drought years or during the dry summers, the risk of forest fires has escalated in recent decades. On August 23, 2013, over a 24-h period approximately 3000 lightning strikes occurred in northern California sparking 98 new fires. Earlier on August 19, 2013, there were over 25,400 lightning strikes in California. 100 confirmed new fires sparked burning over 5000 acres.

California big trees show great fires as far back as 245 CE, and again in 1441, 1580, and 1797. Among spectacular fires, the Matilija Canyon fire in 1932 on the Santa Barbara National Forest spread over an area 32 miles long and eight miles wide, over 220,000 acres.

Of the top 20 largest California fires, eight were lightning caused, including the Rush Fire, 2012, near Lassen that burned over 270,000 acres slightly behind the largest fire ever. However, each year lightning triggers numerous wildfires that can be devastating to large areas of the state. In 2008, lightning touched off more than 100 fires that resulted in more than a billion dollars in damage to forests and property. The smoke from the fires was so severe that in June, officials canceled the Western States Endurance Run, a 100-mile-long ultramarathon that takes runners from the base of the Squaw Valley ski resort in the Sierra Nevada to the Placer High School track in Auburn, California.

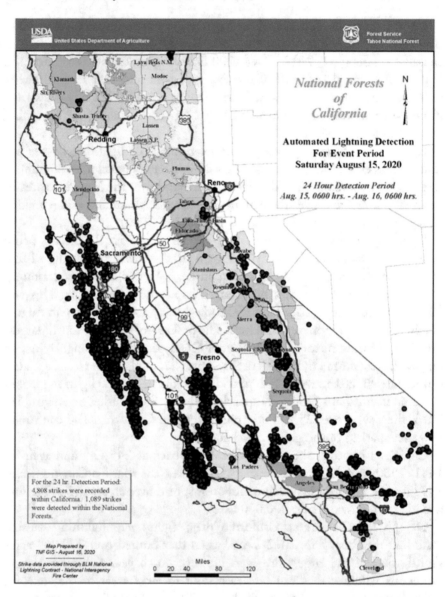

Fig. 10.4 Summer thunderstorms produced 1089 lightning strikes on August 15, 2020. Most lightning struck in northern California, where wildland fires caused record damage (National Interagency Fire Center, posted on YubaNet.com on August 16, 2020)

Record heat and drought in 2020 and 2021 led to the worst fire seasons in California history. In August 2020, lightning that accompanies dry thunderstorms resulted in a series of wildfires in northern California. The unusual number of these storms resulted from very hot and dry air near Earth's surface coupled with moisture from the remains of Tropical Storm Fausto that had moved into the San Francisco Bay area. Between 1.5 and 2.1 million acres were burned. In 2021, the fire season began earlier and came in second only to 2020 in number of acres burned.

According to NOAA's Storm Event database, in the 20-year period spanning 2002 to 2022, there were 157 days with lightning and a total of 256 events. Ten people were killed during this period, and 18 others injured. There were 92 days with property damage and 11 days with crop damage.

10.3 Tornadoes

Tornadoes in California are relatively rare but not nearly as rare as one may think. NOAA's Storm Events Database shows that California experiences, on average, eight tornadoes per year. However, these tornadoes tend to be on the weak side.

Tornadoes are categorized by the Enhanced Fujita scale and range from EF0 to EF5. This revision of the earlier Fujita scale (F-scale) of F0 to F5 was made to better account for damage categories and a better understanding of wind speeds in tornadoes. The original F-scale, likely overestimated wind speeds in many cases. The winds in an EF0 range from 65 to 85 mph but exceed 200 mph in an EF5. The vast majority of California tornadoes have been in the EF0 and EF1 categories or F0 to F1.

Tornadoes have a long history in California, with over 400 confirmed touchdowns in the golden state since 1950. Most of these have come since the 1980s as observations, Doppler radar and volunteer spotter programs account for increasing reporting each decade. Over 250 tornadoes were reported from 1998 to 2012. Ninety-nine of those were spotted in the Los Angeles Basin. The 1997–98 El Niño event showed 28 California tornadoes. In an oddity of nature, 2005 saw more tornadoes in California (30) than in Oklahoma!

More tornado reports occur in the more populous Los Angeles Basin, which is known to be "the tornado capital of the West" having the most tornadoes per square area than anywhere west of the Rockies (Fig. 10.5). Monthly occurrence of tornadoes reaches a maximum in March with the winter storm season. A secondary maximum associated with the North American Monsoon occurs in August (Fig. 10.6).

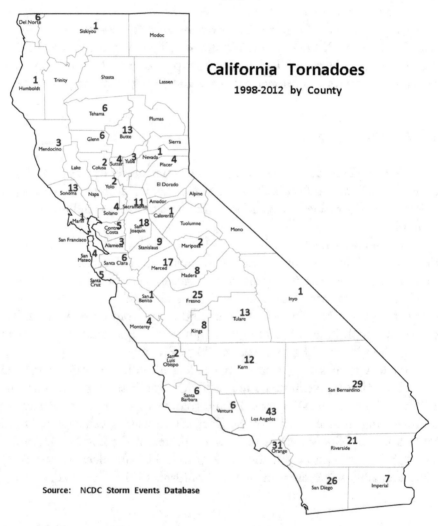

Fig. 10.5 California tornadoes, 1998–2012, by county (NCDC, goldengateweather. com, courtesy Jan Null)

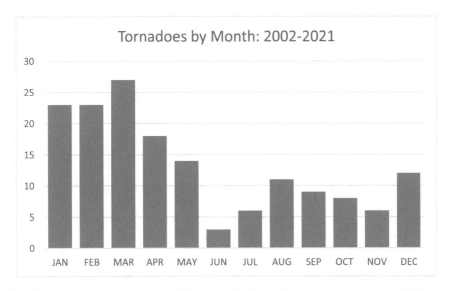

Fig. 10.6 Tornadoes by month (data from NOAA's Storm Events Database, NCEI)

10.3.1 Significant California Tornadoes

An F2 (113–157 mph) tornado in Santa Clara County, with a path length of 5.7 miles did result in 2.5 million dollars in damage on November 11, 1951. An EF2 (111–135 mph) affected March Air Force Base on May 22, 2008, with one injury and 350 thousand dollars in damage. No deaths have ever been associated with tornadoes, although since 1950, there have been 29 injuries.

Only two F3s (wind 158–206 mph) have been recorded and both resulted in significant destruction. On Feb. 2, 1978, one cut a two-mile path through Orange County, causing 2.5 million dollars in property damage and six injuries. The other F3 tornado touched down only briefly in Riverside County (August 16, 1973). There were no injuries, but 25 thousand dollars in damage was reported. Since the new scale was incorporated in 2007, no EF3 (136–165 mph) or higher tornado has been reported.

On Nov. 9, 1982, eight tornadoes were reported in the Los Angeles area in early afternoon, well behind a cold front. This El Niño year storm produced an F2 tornado that damaged the downtown Convention Center.

From 2002 through 2021, only one EF2 tornado was reported. This occurred in Glen and Butte Counties in May of 2011. Damage was estimated at 120,000 dollars (Figs. 10.7 and 10.8).

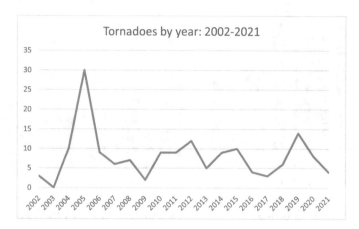

Fig. 10.7 Tornado occurrence by year from 2002 through 2021 ranges from zero in 2003 to 30 in 2005 (data from NOAA's Storm Events Database, NCEI)

Fig. 10.8 Tornado over San Nicolas Island (Courtesy Bill Reid)

10.4 Waterspouts

A waterspout is a vortex that extends from the base of cumulus clouds to a water surface. Waterspouts can be either "fair weather" or "tornadic." Fair weather waterspouts are usually associated with cumulus clouds over a large body of water on warm summer days. They usually move very little, but if they move over land, they dissipate. They rarely cause significant damage but may be a hazard to small watercraft that sail near them. Tornadic waterspouts, on the other hand, are simply tornadoes that form over or move over water. Unlike the fair-weather type, these waterspouts are associated with severe thunderstorms and can result in a great deal of damage.

Waterspout sightings occur almost exclusively along the coast with the exceptions of one that was reported along the Sacramento River and those reported in the San Francisco Bay. However, on September 26, 1998, according to NOAA's Storm Events Database, 11 waterspouts were reported on the north shore of Lake Tahoe. Other lakes may experience waterspouts as well, but reports are rare. There may be a population bias in the frequency of reports as the south coast, especially south of Point Conception has the majority of events. Most northern reports are generally near urban centers, especially the San Francisco Peninsula. Another possible result of population bias is the fact that all waterspouts are reported during the daylight hours. This may also be explained by the fact that waterspouts are associated with fair weather cumulus clouds, that are basically a daytime phenomenon.

Seasonally, waterspouts are rare in summer months when the Pacific High is most dominant. If a waterspout comes onshore, and does not dissipate it would be a tornadic waterspout and be reported as a tornado. This has happened occasionally, especially in the Los Angeles and Orange Counties (see photo of Huntington Beach waterspout, Fig. 10.9). An unusual tornadic waterspout that became a tornado over land showered a number of fish on the city of Chula Vista in San Diego County April 13, 1956.

NOAA's Storm Events Database only has limited waterspout data (Fig. 10.10).

10.5 Funnel Clouds

Funnel clouds are vortices that drop from thunderstorm cumulonimbus clouds but, unlike tornadoes, do not touch the ground. Damage is rare from funnel clouds, though some are invisible near the ground until they kick up debris. Funnel cloud reports would again be reported when seen

Fig. 10.9 Waterspout off Huntington Beach 02/19/05 (This photo of Huntington Beach is courtesy of Tripadvisor.com)

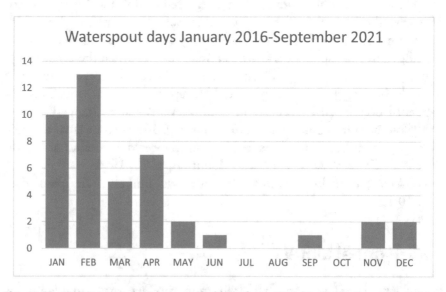

Fig. 10.10 Waterspout days by month (data from NOAA's Storm Events Database, NCEI)

by observers, therefore no nocturnal reports. Seasonally funnel clouds are reported throughout the year, but far fewer in early summer than the other seasons. Reports of funnel clouds are as frequent as tornadoes and show similar distribution over the state (Fig. 10.5). Compared with tornadoes, the proportion of total funnel cloud events is higher in the Central Valley. There are also reports of funnel clouds in mountainous areas near Lake Tahoe and to the far northeast, while no tornadoes are reported in these areas. Population bias appears to be present here as well as with tornadoes as more sightings occur in heavily populated areas. With a similar distribution, 170 funnel clouds were reported over California from 2002 through 2021. This compares to 160 tornadoes for the same period.

10.6   Hail

Hail is a result of strong updrafts in a thunderstorm. The stronger the updraft, the larger hailstones can grow. When a hailstone grows large enough to overcome the thunderstorm updraft, it falls to the ground. Only a few cases of reported damaging hail meet the criteria of "severe hail" (0.75 in. in diameter or greater-National Weather Service). These larger sized hailstones can cause considerable damage to crops, property and injuries to people.

In the period 2012 through 2021, 348 hail events were reported occurring over 159 days with some days having several events. Figure 10.11 shows the number of hail events by month. Similar to tornadoes, hail shows a maximum occurrence in March. A secondary peak occurs in July associated with the North American monsoon (Fig. 10.12).

Hail shows quite a bit of year-to-year variability. Figure 10.11 shows a rise that begins in 2016, peaking in 2018.

Based on reported events in the NOAA's Storm Events Database, damaging hail can occur throughout the state. Reports in the heavily populated coast are quite sparse, while the Central Valley shows the largest concentration of reports, with the greatest clustering in Fresno County. Since this is a densely farmed region, these reports may reflect crop and property damage. A majority of events seems to correspond to Interstate 99 and I-5 where the largest cities are also found. Eastern regions of the state show few reports, but that may be due to the smaller populations in these rural areas. If hail occurs

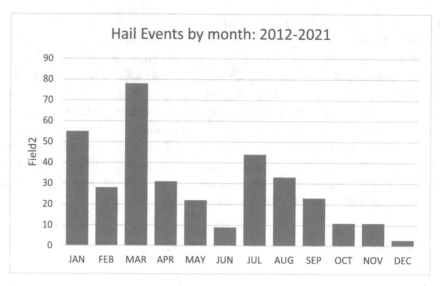

Fig. 10.11 Hail events by month (data from NOAA's Storm Events Database, NCEI)

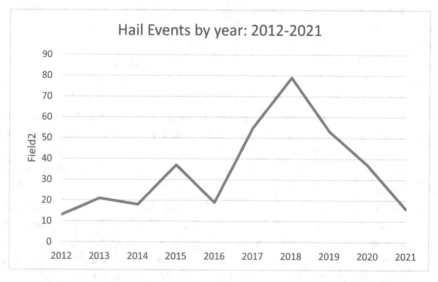

Fig. 10.12 Hail events by year (data from NOAA's Storm Events Database, NCEI)

in uninhabited regions where no damage occurs, the chance of a report being made is less than where you have sensitive property (glass windows, greenhouses) or crops during the growing season. Another factor in hail frequency points to population increases, when comparing five-year periods, there were 42 events reported in 1990–94 jumping to 162 events in 1995–99.

The largest reported hailstone, according to the National Weather Service office in Sacramento, occurred near Corning, California on Jan. 23, 2016. It was a star-shaped hailstone measuring three inches in diameter from opposite ends of the star, tying the state record hailstone size, which was first set on September 2, 1960, in San Diego County. On May 24, 2019, hail with diameters of three inches was reported near Redding. Estimates of damage to cars, mainly rear windows and windshields were at $7.5 million.

NOAA's database indicated there had been only 16 instances of hail greater than golf-ball size—1.75 in. in diameter—from 1950 through December 31, 2021. The most recent occurrence was more than ten years ago in Siskiyou County on July 26, 2010. In another case, widespread hail covered much of Sacramento on February 25, 2018, causing traffic delays. On April 15, 2018, hail hit the Bay area closing schools in Oakland, then covered parts of Sacramento with more hail.

Crop damage totaled nearly $200 million since 1950. Severe hailstorm took place on March 8 and 10, 1986, in Fresno and Tulare counties that left $40 million in damage. A report on crop insurance losses for 1948–2001 has Fresno County with the highest claimed losses in crops, with nearly half of the state total. Largest losses were to nectarines and plums.

It may have resembled snow, but on November 12, 2003, a foot of hail fell on Watts and south-central Los Angeles, with numerous power outages from frequent lightning. National Weather Service officials said that 5.31 in. of rain fell at 96th Street and Central Avenue in Los Angeles in less than three hours. Motorists were stranded in flooded intersections during rush hour. Damage totaled $3.5 million. The storm featured a nearly stationary cumulonimbus cloud.

10.7   Flash Floods

Heavy thunderstorm rain can cause extensive damage and threaten lives. The difference between flash floods and other floods is timing. While most floods depend on soil conditions and rainfall amounts and duration, flash flooding may occur within six hours of the rain event and is dependent on rainfall and duration and are usually associated with thunderstorms.

Coastal plains, foothills, and valleys show a winter tendency to flash flood events, while inland southern California mountains and deserts experience more flash flooding events in summer.

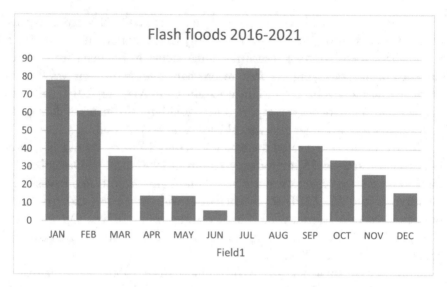

Fig. 10.13 Flash floods by month (data from NOAA's Storm Events Database, NCEI)

Flash floods occur with greater frequency than other severe events. Looking at just the five-year period from 2016 through 2021, there 473 events over 147 days. This resulted in 10 deaths, 16 injuries, and nearly $180 million in damage. The flood with the most damage occurred on Mar 22, 2018, in Tuolumne County with total damage at 43 million dollars. Moccasin Dam nearly failed, and there was much damage to Highways 49 and 132 (NOAA's Storm Events Database) (Fig. 10.13).

Bibliography

Aldrich, J. H., & Meadows, M. (1956). *Southland weather handbook*. The Weather Press.

Blier, W., & Battan, K. A. (1994). On the incidence of tornadoes in California. *Weather and Forecasting, 9,* 301–315.

Brown, J. (2005). *Severe weather in southern California* (MA Thesis), California State University, Northridge.

Cal Fires. (2022). *Incidents.* https://www.fire.ca.gov/incidents/

Grotjahn, R. (2000). Multiple waterspouts at Lake Tahoe. *Bulletin of the American Meteorological Society, 81,* 695–702. https://www.jstor.org/stable/26215132#met adata_info_tab_contents

Guthrie, J. D. (1936). *Great forest fires of America*. US Department of Agriculture, Forest Service.

Hales, J. E. (1985). Synoptic features associated with Los Angeles tornado occurrences. *Bulletin of the American Meteorological Society, 66*, 657–662.

Halvorson, D. A. (1971). *Tornado and funnel clouds in San Diego County.* Western Region Technical Attachment No. 71-33.

Lipari, G. S., & Monteverdi, J. P. (2000). Convective and shear parameters associated with northern and central California tornadoes during the period 1990–94. In *Preprints—20th American Meteorological Society Conference on Severe Local Storms.* Orlando, Florida.

Los Angeles Department of Public Works. (2003). *Flood control system evaluation report.* November 12, 2003 storm. https://file.lacounty.gov/SDSInter/bos/bc/014145_FLOODCONTROLSYSTEM.pdf

Meier, J., & Thompson, R. (2009). *Lightning climatology for the WFO Los Angeles/Oxnard California warning area.* https://www.weather.gov/media/wrh/online_publications/talite/talite1101.pdf

Monteverdi, J. P., Braun, S. A., & Trimble, T. C. (1988). Funnel clouds in the San Joaquin Valley, California. *Monthly Weather Review, 116*, 782–789.

Nakamura, K. (1987). California's valley and high country climate. *American Weather Observer, 4*(1), 9–10.

National Centers for Environmental Information. (2022). *Storm events database.* https://www.ncdc.noaa.gov/stormevents/

National Interagency Fire Center. (2022). *Lightning-caused and human-caused fires and acres.* https://www.nifc.gov/sites/default/files/document-media/Lightning-Human-Caused_2.pdf

National Oceanic and Atmospheric Administration (NOAA). JetStream. *Introduction to thunderstorms.* https://www.weather.gov/jetstream/tstorms_intro

National Oceanic and Atmospheric Administration (NOAA). *JetStream thunderstorm hazards-hail.* https://www.weather.gov/jetstream/hail

National Weather Service. (2017). *A history of significant weather events in southern California, Organized by weather type.* https://www.weather.gov/media/sgx/documents/weatherhistory.pdf

Neyman, I. (2013). *Forecasting California thunderstorms* (MA Thesis), California State University, Northridge.

Null, J. (2022, January 20). *Tornado statistics.* Golden Gate Weather Services. https://ggweather.com/ca_tornado.htm

Null, J., & Mogil, M. (2010). The weather and climate of California. *Weatherwise, 63*, 16–23.

SF Gate. (2017, September 11). *A rare bay area lightning storm.* https://www.sfgate.com/local-donotuse/slideshow/A-rare-Bay-Area-lightning-storm-9-11-17-163644.php

Small, I. J. (1999). An observational study of Island effect bands: Precipitation producers in Southern California. Western Region Technical Attachment No. 99-18, 11 p (Available from the National Weather Service, 125 S. State St., Salt Lake City, UT 84138).

Tripadvisor. (2022). Huntington beach. https://www.tripadvisor.com/Tourism-g32
    513-Huntington_Beach_California-Vacations.html
YubaNet.com. (2022). *Fire News*. https://yubanet.com/fires/

11

Climate Change in California: Past, Present, and Future

Fig. 11.1 National Guard and other agencies were not able to contain the fast-moving Camp Fire (California National Guard)

© The Author(s), under exclusive license to Springer Nature
Switzerland AG 2023
S. LaDochy and M. Witiw, *Fire and Rain*,
https://doi.org/10.1007/978-3-031-32273-0_11

The future ain't what it used to be.
—*Yogi Berra*

Recent droughts and rising temperatures have fueled the increasing incidents of California wildfires. The most deadly and costliest wildfire in state history was the Camp Fire in November 2018. An estimated 88 deaths occurred while total costs of destruction were estimated at $16.5 billion by Munich Re, a reinsurance firm, making it the world's costliest disaster that year. Normally, winter rains would have prevented such an event, but a lingering drought set the stage for tragedy. Instead of late summer lightning, the culprit was a faulty utility line. Strong easterly winds quickly swept flames across several towns, including Paradise, which was almost entirely destroyed. Climate change may not have caused the Camp Fire event but continued warming and frequent droughts may be leading to more tragedies like Paradise.

11.1   What Is Climate Change?

Over the past few years, we have seen headlines like more and stronger hurricanes; record snow in Philadelphia and Boston; drought in California. These are all signs of climate change—or are they? The fact is that climate is always changing, but what is the difference now?

Climate change is the altering of the state of the climate to one that is significantly different over a lengthy period of time, usually decades or longer. Our environment, and our climate, is changing. There are many natural inputs into the climate system such as volcanic activity as well as solar and astronomical cycles. But today, there is an increasing human input that affects Earth's energy budget and likely climate. These include increases in greenhouse gases and deforestation. Changes in our atmosphere and on Earth's surface continue to alter the global distribution of heating, forcing our climate to change (Fig. 11.1).

A shorter feature of our climate is its variability. This variability ranges over multiple time and space scales and includes phenomena such as El Niño/La Niña, which can lead to floods and droughts and even multi-year changes in temperature and precipitation patterns. Some examples of longer timescale variabilities might include a series of abnormally mild or exceptionally severe winters and/or summers. Climate has been changing globally and locally throughout the evolution of our planet. California's climate has also evolved over time with periods of glaciation, megadroughts, massive floods, and increasing heat waves. Evidence of these changes has come from fossils, tree rings, speleothems (depositions in caves like stalagmites and stalactites), pollen, middens (refuse heaps), scat (animal waste) and sediment layers, among others. Here, we will briefly look at climate change as it pertains to present weather and climate conditions. By looking at the past, we can determine how fast change occurs and how different these changes were compared with today's environment. Viewing historical records, we see the frequency and severity of extreme events. Seeing the most recent trends can prepare us for possible future climatic changes. First, we start with a brief look at California's prehistorical climates.

11.2 Prehistorical Climate Change

California's prehistorical climate mirrors that of Earth's. During its early years from about 4.6 billion to 2.3 billion years ago, Earth was very warm. This was largely due to the presence of greenhouse gases like methane and carbon dioxide as well as geothermal heat from Earth's interior that was gradually cooling since the creation of Earth. Around 2.3 billion years ago, oxygen levels rose rapidly being produced by photosynthesis from the increasing plant life. With the reduction of greenhouse gases, Earth quickly cooled. The result of this was "Snowball Earth"—the first of four icehouses, each lasting 10s of millions of years. We are currently in the Fourth Icehouse which began about 35 million years ago. During icehouses glaciers periodically advance and then recede. Within the Fourth Icehouse, we have had what are termed the Pleistocene Glaciations which lasted from about 1.8 million to 11,500 years ago. During this time, there were periodic advances of northern hemisphere mid-latitude ice sheets. It is widely believed that these advances (and retreats) are associated with variations in Earth's orbit. These variations are called the Milankovitch cycles after the twentieth century scientist who discovered them (Fig. 11.2).

The eccentricity cycle is about 100,000 years. Earth goes from a nearly circular orbit to one that is much more elliptical. The precession cycle lasts about 20,000 years. Currently, Earth is closest to the sun in January. In about 10,000 years, it will be closest in July. The obliquity cycle is about 41,000 years and involves the tilt of Earth's axis. Currently, the axis is tilted at 23.5°. This tilt varies between 22° and 24.5°. Collectively, it is widely believed that these cycles are responsible for glacial advances and retreats with major advances occurring every 100,000 years and minor advances between 20,000 and 25,000 years. Currently, we are in an interglacial period.

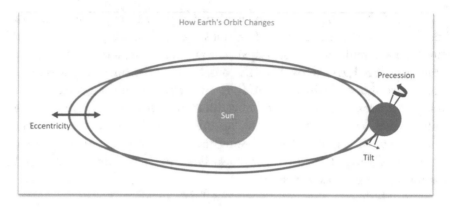

Fig. 11.2   Three Milankovitch cycles (Graphic by Harrison Blizzard)

11.3   Prehistorical Climate Change in California

The cooling of deep seas and upwelling of cooling water along the Pacific coast resulted in the development of summer drought during the Miocene Epoch (more than 5 million years ago) a situation that still prevails today. California's climate has fluctuated with glacial–interglacial cycles since the Pleistocene Epoch (roughly 2.6 million to 11,500 years ago), with precipitation greater than present during glacial advances and climates similar to today's during interglacial periods.

Evidence for cooler wetter climates in California includes glaciation along the Sierra Nevada's crest and pluvial lakes (which form during wet periods) in the Great Basin. Glaciers were more than 3000 ft below that of modern glaciers in the Sierra Nevada. Pluvial lakes increased beginning 32.0 thousand years ago and reached maximum extents between 15.0 and 13.5 thousand years ago. They then dried or decreased to modern levels by 10 thousand years ago. Lakes in the eastern Sierra Nevada south of Lake Tahoe were shallow and dried out from 5300 to 4800 years ago providing evidence that glaciers in the Sierra Nevada may have been entirely absent during much of the Holocene. This period is coincident with relatively high salinity in San Francisco Bay and dry climate throughout California (The Holocene Epoch began about 12,000 years ago and remains the epoch we are in today).

The early Holocene (12 thousand years ago) may have been wetter than present in far southeastern California. There were several deep lakes in the Lake Mohave basin between eight and 10 thousand years ago, possibly due to enhanced summer monsoons. The mid-Holocene in California (5.0–3.5 thousand years ago) was a period of greater rainfall, enlargement of Sierra Nevada alpine glaciers, and small interior lakes.

From tree ring data, two dry periods were identified to have occurred between 900 and 1400 CE. Dry periods and warm periods appear to be connected. This time is coincident with the medieval warm period chronicled in Europe especially in the Viking chronicles. During the droughts, each lasting as long as a century, California received about 60–70% of today's precipitation. These are now called Medieval Megadroughts. Records also show there have been wet-dry cycles that ranged from 30 to 200 years over the past two millennia.

About 2000 BCE, a cool, wet period occurred. Following this, long droughts interrupted by brief, very wet periods were the rule. In particular, during the period around 950–1150 CE conditions seem to have been unusually dry in many parts of the S.F. watershed, reducing the freshwater flows to the estuary, and shifting tidal marsh plants to less diverse, but more salt-tolerant plants. During the Little Ice Age (around 1300–1150 CE), the climate was unusually wet and cool, with salinity in San Francisco Bay decreasing. Since, about 1850 CE, with the exception of the recent, short-term anomalies, the climate has been relatively stable.

Fire records from several California sources also trace past climates. The Santa Ana winds, and the accompanying raging wildfires have been a part of the ecosystem of the Los Angeles Basin for over 5000 years, dating back to the earliest habitation of the region by the Tongva and Tataviam peoples. Thomas Swetnam from the University of Arizona's tree ring lab studied the fire scars on the giant Sequoias in the Sierra Nevada, and he was able to determine that the fire frequency was much greater during the medieval dry period than today (Brown et al., 1992).

Pollen records of northwest California show that the dense forests have undergone great changes in composition and structure. Sediment cores from Clear Lake in the North Coast Range show that pollen was dominated by pine trees (*Pinus*) in late glacial and oak trees (*Quercus*) during the Holocene period (started about 12,000 years ago). The large shift toward *Quercus* began at around 15 thousand years ago. Today, the region is dominated by oak woodland.

11.4   More Recent Climate Change

In the past millennium, we can see that longer term climate change appears to be associated with solar activity such as the Little Ice Age. Shorter changes occur after major volcanic eruptions. Tambora in 1816, Krakatoa in 1883, and Pinatubo in 1991 were all followed by a year or two of global cooling.

Fig. 11.3 Sunspots (or lack of them) as shown here have been thought to have an effect on climate (NASA Solar Dynamics Laboratory)

11.4.1   The Little Ice Age

The "Little Ice Age" was a period of significantly cooler temperatures that lasted from about 1300 CE to around 1850 CE (some references give a slightly longer or slightly shorter period). A period of rapid cooling began before 1600, with temperatures during the remainder of the Little Ice Age being as cold as any period in the Holocene.

It is generally thought the main reason for the Little Ice Age was decreased solar activity as evidenced by fewer sunspots. Coldest temperatures occurred during the Maunder minimum of sunspot activity (1645 CE–1715 CE). The Little Ice Age is well documented in European chronicles. Although there is little in the way of historical accounts for California, proxy data (such as tree rings and fossils) do indicate California also experienced cold temperatures and glacial expansion (Fig. 11.3).

11.4.2   Volcanoes

When there is a major volcano eruption, sulfur dioxide can be transported into the stratosphere. When this happens, droplets of sulfuric acid form that reflect incoming solar radiation, resulting in cooling at Earth's surface. After the eruption of Tambora in 1816, we saw the Year Without a Summer, evidenced by late frosts and snowfall throughout much of the northern United States. A similar cooling occurred after Krakatoa (maybe related to the snow in San Francisco in the 1880s?). When Mt. Pinatubo erupted in 1991, the result was worldwide cooling of about 0.5 °C.

The following winter, temperatures in California varied from near normal in the north to up to 1 °C below normal in the south. The summer of 1992 was up to 1 °C below normal, with spectacular red sunsets from the high-altitude ash. The cooling from the spread of Pinatubo's sulfuric acid cloud was seen in both hemispheres for up to two years. The averaged temperature changes that occurred after the eruption of Mt. Pinatubo can be seen in Fig. 11.4.

11.5 Missionary Records/Early Historical

Early written reports of California weather came from missions, diaries, and tree ring data going back to about 1700. These proxies to instrumental data add length to our weather records.

Diaries of early Spaniards describe weather, crops, rainfall, and stream flow since 1769 mostly in southern California, but also in the San Francisco Bay area. Other studies describe floods and droughts from 1769 to 1834 using mission records, diaries (weather and agriculture), and tree ring data for southern California (back to 1700).

Numerous floods were described between the years of 1811 and 1884 and occurred about every 10 years on average. Notably among them was the flood of 1850 when the valley of the Sacramento River, and the city of Sacramento became a sea of gondolas, when gold rush miners navigated wagon-boxes, bakers' troughs, crockery crates, and whisky-kegs. The flood of 1851 and 1852 followed and brought disaster and financial ruin to miners near the camps along the Sierras. Dr. W. F. Edgar, surgeon of Fort Miller near the headwaters of the San Joaquin recorded 46 in. of water that fell during the months of January and February 1852 (Guinn, 1890).

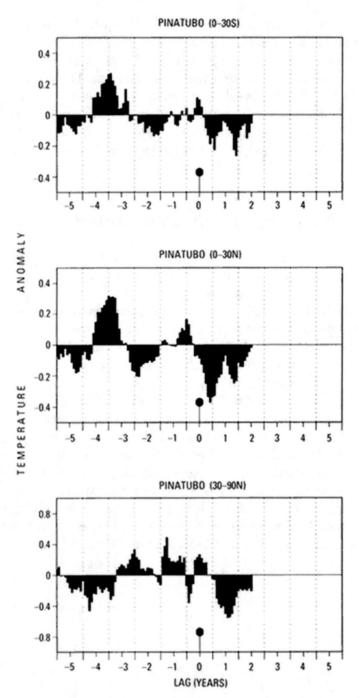

Fig. 11.4 Averaged temperature change from normal (anomaly) is shown at various latitudes for years before and after the eruption of Mt. Pinatubo (Self et al., 1993). Note that temperatures cooled for up to two years after the eruption

Conversely, droughts were also frequent. A significant one occurred from fall 1862 and lasted until the winter of 1864–1865. Less than four inches of rain fell in both the winters of 1862–63 and 1863–64 through most of the populated areas of the state.

Although floods often cause considerable damage, they also have beneficial effects. They fill up the springs and mountain lakes and reservoirs that feed supply water for irrigation during the long dry season. A flood year is almost always followed by good harvests. Increased facilities for irrigation, artesian wells, the building of reservoirs for water storage, and the more economic use of water, have done much to counter the adverse effects of the dreaded dry year.

What about *REALLY DRY?* The San Luis Rey de Francia Mission in north San Diego County, near Oceanside, had a Mission observation record of zero precipitation (not even a trace) from May 1790 until November 1791! The July 1790–June 1791 rain season in Los Angeles brought less than one-half inch (one report of only 0.08 in.), but none of those pitiful "storms" ever made it to San Diego County. The Mission in the town of San Diego (not yet a city) did record small amounts of rain in early October 1790 and August 1791 (probably monsoon or tropical storm rainfall). None of it reached the San Luis Rey Mission.

Two Years Before the Mast (1840):

In his book, *Two Years Before the Mast*, Richard Henry Dana Jr. writes about his visits to coastal southern California and San Francisco between 1834 and 1836. His descriptions of weather include strong offshore winds, mild winters, and devastating fires. See Appendix C for two excerpts from the book.

11.6 Current Climate Change

For California, "*The past 150 years have been unusually wet when viewed over the past 2000 years, so the twentieth century was a wetter century*" (Source: Warnert, University of California Division of Agriculture and Natural Resources: Green Blog.)

Many reconstructions suggest that compared to the past notably stable conditions have prevailed over the instrumental period, after about 1850 CE, even including the severe, short-term anomalies experienced recently.

Recent climate change can be seen from a large number of weather stations scattered throughout the state. However, finding continuous, reliable weather data, is difficult beyond the last 70 years (since 1950) for most stations. Weather stations are more numerous in populated areas, coastal and lower elevated valleys. Much fewer weather stations exist in mountainous regions and in desolate deserts of the southeast and remote northeast. Other complications include station moves (horizontally and/or vertically), changes in instruments (e.g., automated systems in airports) and changes in land use, especially where urbanization or agriculture replace native vegetation.

Recent change in temperature and precipitation has been shown statewide and by climatic divisions, where landscape and land uses are similar among stations. Here, we show some examples of how in the state and individual regions climate has been changing.

11.6.1 Temperature

Figure 11.5 shows the annual average maximum temperatures for the state since 1895. You can see that there is considerable year-to-year variability as well as a general warming over the years. Warming has been accelerated since the 1980s.

The south coast has shown the highest warming trends of the regions, with nearly 3 °F/century, while the north coast barely warms 1 °F/century. Figure 11.6 shows average annual temperatures for more than 120 years up to 2020 for south coast stations. Temperatures since 1895 show periods of warming and cooling throughout the record. These dips and rises correspond to changes in coastal ocean temperatures, such as when there is an El Niño or La Niña. But the general trend shows one of the fastest increases among divisions. Part of the south coast has grown into a densely populated megalopolis. The cause of such rapid warming is likely a combination of general global and state trends and expanding local urban heat islands (see Chap. 4) around cities such as San Diego (Fig. 11.7).

Figure 11.6 shows a 4 °F warming, about twice that of global warming in the same period. Note that there is a great variability from year to year as well as warming and cooling cycles that follow closely coastal ocean temperatures and the PDO (NOAA, National Centers for Environmental Information).

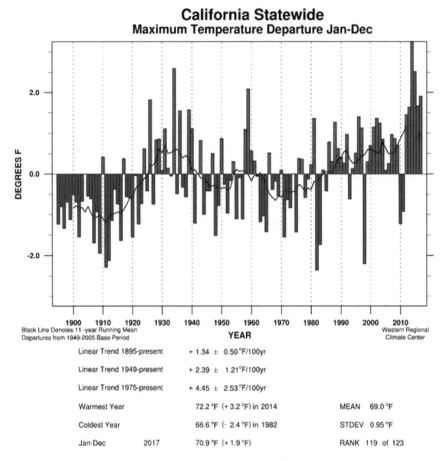

Fig. 11.5 Annual maximum temperature departures from the long-term average for the state from 1895 show that since the 1980s, there has been a warming trend (California Climate Tracker, WRCC)

On the other hand, cool weather is disappearing in the state. Chill hours, (see Fig. 11.8) or hours where the temperature is below 50 °F, are important for many fruit and nut crops. Number of chill hours has been quite variable over the past 60 years but has shown a notable decrease since 1990.

Based on the moderate B1 (lower emissions) scenario, future estimates show a definite downward trend. The number of chilling hours at the end of this century in Davis is expected to be half or less than during the 1980s. For the B1 scenario, many crops, such as pears and pistachios, will not be sustainable in many currently farmed areas of California (see also Chap. 13 Agriculture).

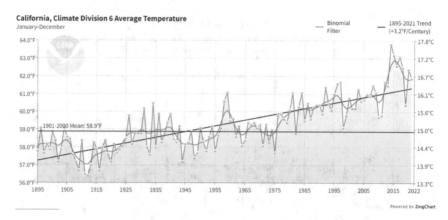

Fig. 11.6 Over a century of annual average temperatures in the south coast region (Division 6), NOAA, NCEI data

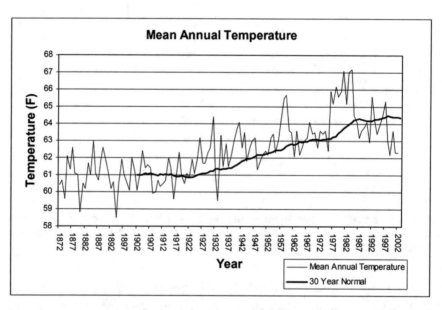

Fig. 11.7 Mean annual temperatures for San Diego show no change from 1872 until 1920, but a warming trend of over 3 °F since. The thick solid line smooths the variability by using a moving 30-year average (NWS Western Regional Headquarters: Evans and Halvorson, Climate of San Diego 1998)

11.6.2 Precipitation

California recently experienced its driest five years (2014–2018) in recorded history (since 1895), and one of the longest dry periods, which started in 2000 and is now considered a megadrought. Though precipitation appeared

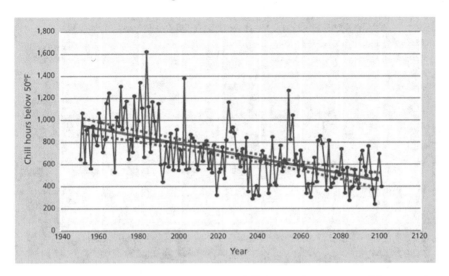

Fig. 11.8 Past and projected annual total chill hours for Davis, California values through 2003 are observations; values after are based upon average projections of two climate models driven by moderate B1 (lower emissions) scenario. Adapted from Baldocchi and Wong (2008). Reproduced with permission from Springer Nature

to be increasing throughout the state during earlier decades of the last century (Fig. 11.11), in recent decades (since 1975) northern and central California have become wetter, while southern California has become drier.

Figure 11.9 shows the precipitation patterns for the state since 1950. Since temperatures affect evapotranspiration, higher temperatures lead to greater water loss or increasing drought (Fig. 11.10). The recent droughts show increasing severity with short breaks (wet spells). Droughts have been especially frequent in southern California since 2000 (Fig. 11.11).

The US Drought Monitor began in 2000. Figure 11.12 shows the percent of the state in drought since 2000. The longest duration of drought (D1–D4) in California lasted 376 weeks from December 27, 2011, until March 5, 2019. The most intense period of drought was 2014–2017 when over 50% of the state reached D4 (NCDC, NOAA). D0—precursor to drought, D1—moderate, D2—severe, D3—extreme, D4—exceptional are categories used by the Monitor.

Snow amounts have also varied greatly in recent decades following wet and dry periods generally. However, the snow season has been shifting toward earlier runoff in recent decades attributed to (1) more precipitation falling as rain instead of snow and (2) earlier snowmelt. The water content at the end of winter (April) snowpack has also been decreasing. A recent study published in the journal *Water* suggests the snow line has risen about 1200 ft in the

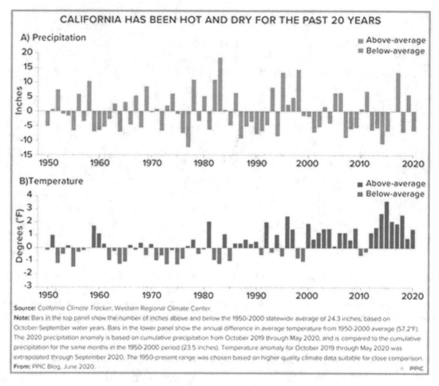

Fig. 11.9 While temperatures in the state have been climbing, precipitation goes through wet and dry periods. Note that since 2000, most years were dry and very warm compared to the long-term averages (Public Policy Institute of California; data from WRCC, California Climate Tracker)

northern Sierra Nevada due to rising temperatures since 2007 (Dettinger et al., 2011). The January–March 2022 snowfall in the Sierras was the lowest ever recorded.

Other recent changes show that fog and low clouds have become less frequent, especially along the south coast. On the other hand, total moisture (specific humidity) in the warmer air seems to be increasing as Pacific Ocean temperatures rise and evapotranspiration rises as well.

Table 11.1 shows how climate has changed by comparing data at three locations from 1931–1960 to 1991–2020. Airports in the table are San Francisco and Los Angeles International and Sacramento Municipal/Executive (name change in 1967). Although municipal locations show similar changes, they tend to relocate from different downtown locations so fixed airport locations were used in the table.

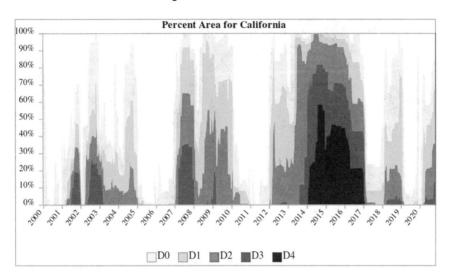

Fig. 11.10 Percent of the state in drought is shown from 2000. *Source* Mount and Dettinger (2020)

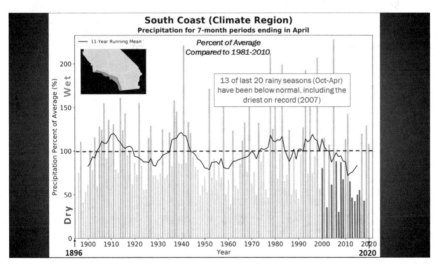

Fig. 11.11 In respects to the long-term record, the populous south coast has had its share of droughts recently with most of the rainy seasons since the 1997–98 El Niño being below normal (Gomberg, 2021)

In the table an increase in temperature can be noted at all locations, most notably in January. Conversely, while northern California locations showed an increase in precipitation, Los Angeles showed a decrease (NOAA's National Centers for Environmental Information).

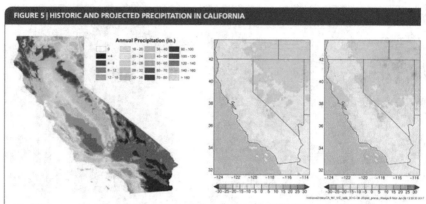

FIGURE 5 | HISTORIC AND PROJECTED PRECIPITATION IN CALIFORNIA

Left panel: Average annual precipitation in California. Right two panels: Projected percent changes (2070-2100 relative to 1950-2005) of annual precipitation, averaged over 10 LOCA downscaled GCMs selected for the Fourth Assessment for RCP 4.5 (left) and RCP 8.5 (right) scenarios. Sources: Left: modified from PRISM, 2018; Right: Pierce et al., 2018.

Fourth Climate Change Assessment   Statewide Summary Report | 25

Fig. 11.12 Projected precipitation maps show possible future drying under two warming scenarios. For higher GHGs (far right), drying is more severe, especially in southern California, while northern California may become wetter (California Energy Commission: California Fourth Climate Change Assessment)

Table 11.1 Temperature and precipitation comparison

Location	San Francisco		Los Angeles		Sacramento	
Years	1931–1960	1991–2020	1931–1960	1991–2020	1931–1960	1991–2020
Average January temperature	48.4	51.3	54.4	57.9	45.2	47.6
Average July temperature	62.7	64.0	69.1	69.6	75.4	75.9
Annual precipitation	18.69	19.64	14.68	12.23	16.29	18.14

11.7 Future Climate Change

What will California's weather and climate be like in the future? Will the mild, Mediterranean climate continue to attract immigrants and tourists to this perceived paradise? By looking at the past and present, scientists can speculate about what the future might bring. California's Fourth Climate Change Assessment (California Energy Commission, 2018) includes over forty-four technical peer-reviewed reports that examine specific aspects of climate change in California, including projections of future climate change impacts, analysis of vulnerabilities and adaptation strategies for various sectors, and social and governance solutions for climate adaptation.

11.7.1 Temperature

Like global climate projections, California will likely continue to warm with more record high temperatures broken and less minimum temperatures broken in the coming decades. Heat waves are also projected to continue increasing both in frequency and duration. Heat-related events will lead to more hospitalizations and deaths, not only from excess heat but by worsening air pollution caused by higher temperatures. Higher temperatures will also lead to greater energy demand (air conditioning for example) and need for water. How much warmer? This depends mainly on future greenhouse gases (GHG). Climate models which consider emissions to strongly rise through 2050 and plateau around 2100 project an increase of 8–10 °F by 2100 Other scenarios that use a less extreme GHG increase yield somewhat lower projected increases.

11.7.2 Precipitation

In a warmer world, the hydrological cycle is accelerated. There will be more evaporation and potentially more moisture available for condensation and precipitation. Predictions tend to favor wetter regions getting wetter, while drier areas becoming drier. The United States southwest is predicted to become hotter and drier, while higher latitudes become wetter. In California, this would lead to a drier southern portion, while central and northern regions would have higher precipitation (Fig. 11.12). More of this precipitation would be in the form of rain rather than snow. Snowfall would not only

Historical and Projected California Snowpack

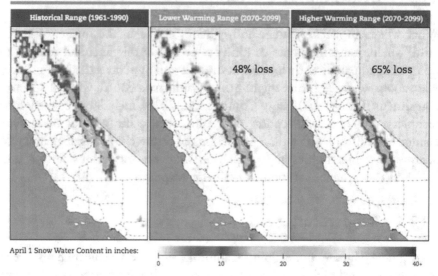

Fig. 11.13 Projected snowpack amounts under different GHG scenarios show future snowpack losses (California Energy Commission: California Fourth Climate Change Assessment)

be less in general but would melt early in the spring (Fig. 11.13). This would leave less water resources behind reservoirs for the hot, summer months. Greater runoff of winter rains may mean greater risk of flooding.

11.7.3 Fires, Agriculture

Climate change will make forests more susceptible to wildland fires in the future. Each year, the number of acres burned has increased, with 2020 breaking the state's record for most forest acreage burned. If greenhouse gases continue to increase, projections call for a 77% average increase in forest acres burned by the end of the century. Agriculture may also change or adapt to warmer, wetter or warmer, and drier conditions. Some crops may have to shift geographically to cooler regions, while others may even benefit from less risk from frost. Many climate scientists, however, contend that the twentieth century was unusually stable and that the coming decades may exhibit more extremes.

Bibliography

Adam, D. P., & West, G. J. (1983). Temperature and precipitation estimates through the last glacial cycle from Clear Lake, California pollen data. *Science, 219*, 168–170.

Baldocchi, D., & Wong, S. (2008). Accumulated, winter chill is decreasing in the fruit growing regions of California. *Climate Change, 87*, 153–166. https://doi.org/10.1007/s10584-007-9367-8

Barnett, T. P., Adam, J. C., & Lettenmaier, D. P. (2005). Potential impacts of a warming climate on water availability in snow-dominated regions. *Nature, 438*, 303–309.

Berghuijs, W. R., Woods, R. A., & Hrachowitz, M. (2014). A precipitation shift from snow towards rain leads to a decrease in streamflow. *Nature Climate Change, 4*, 583–586.

Brown, P. M., Hughes, M. K., Baisan, C. H., Swetnam, T. W., & Caprio, A. C. (1992). Giant sequoia ring-width chronologies from the central Sierra Nevada, California. *Tree-Ring Bulletin, 52*, 1–14.

California Air Resources Board. (2018). *2018 progress report: California's sustainable communities and climate protection act*. https://ww2.arb.ca.gov/sites/default/files/2018-11/Final2018Report_SB150_112618_02_Report.pdf

California Climate Tracker. (2022). *State climate trackers*. Western Regional Climate Center, Desert Research Institute. https://wrcc.dri.edu/my/climate/tracker

California Department of Water Resources, Sacramento. (1980). *California rainfall summary. Monthly total precipitation*. 1849–1984.

California National Guard. (2022). https://calguard.ca.gov/

California's Energy Commission. (2018). *California's fourth climate change assessment—California's changing climate 2018*. https://www.energy.ca.gov/sites/default/files/2019-11/20180827_Summary_Brochure_ADA.pdf

Cayan, D. R. (1996). Interannual climatic variability and snowpack in the western United States. *Journal of Climate, 9*, 928–948.

Cayan, D. R, Bromirski, P. D., Hayhoe, K., Tyree, M., Dettinger, M. D., & Flick, R. E. (2008a). Climate change projections of sea level extremes along the California coast. *Climatic Change, 87*, 57–73. http://doi.org/10.1007/s10584-007-9376-7

Cayan, D. R., Maurer, E. P., Dettinger, M. D., Tyree, M., & Hayhoe, K. (2008b). Climate change scenarios for the California region. *Climatic Change, 87*, 21–42. http://doi.org/10.1007/s10584-007-9377-6

Connor, G. (2008). *History of weather observations San Diego, California, 1849–1948*. Midwestern Regional Climate Center. https://mrcc.illinois.edu/FORTS/histories/CA_San_Diego_Conner.pdf

Cook, B. I., Ault, T. R., & Smerdon, J. E. (2015). Unprecedented 21st century drought risk in the American Southwest and Central Plains. *Science Advances*, 1e1400082. https://doi.org/10.1126/sciadv.1400082

Dettinger, M. D., Ralph, F. M., Das, T., Neiman, P. J., & Cayan, D. R. (2011). Atmospheric rivers, floods and the water resources of California. *Water, 3*, 445–478. https://www.mdpi.com/2073-4441/3/2/445

Evans, T. E., & Halvorson, D. A. (1998). *The climate of San Diego, California*. NOAA Technical Memorandum, NWS 256, Washington D.C.

Gomberg, D. (2021). *Los Angeles/Oxnard fire weather season review and outlook*. Presentation April 2021. Data: NWS, NOAA.

Granger, O. (1977). Secular fluctuation of seasonal precipitation in lowland California. *Monthly Weather Review, 105*, 386–397.

Guinn, J. M. (1890). *Exceptional years: A history of California floods and droughts* (Vol. 1, No. 5). University of California Press, Historical Society of Southern California. https://www.jstor.org/stable/pdf/41167825.pdf

Hatchett, B. J., Boyle, D. P., Putnam, A. E., & Bassett, S. D. (2015). Placing the 2012–2015 California-Nevada drought into a paleoclimatic context: Insights from Walker Lake, California-Nevada, USA. *Geophysical Research Letters, 42*, 8632–8640.

Hayhoe, K., Cayan, D. R., Field, C. B., & Verville, J. H. (2004). Emissions pathways, climate change and impacts on California. *PNAS, 101*(34). http://doi.org/10.1073/pnas.0404500101

Ingram, B. L., & Malamud-Roam, F. (2015). *West without water: What floods, droughts and other climatic clues tell us about tomorrow*. University of California Press.

Jones, T. L., & Klar, K. (Eds.). (2007). *Prehistory California*. Altamira Press.

Kleeman, M. J. (2008). A preliminary assessment of the sensitivity of air quality in California to global change. *Climate Change, 87*, 273–292. http://doi.org/10.1007/s10584-007-9351-3

Knowles, N., Dettinger, M. D., & Cayan, D. R. (2006). Trends in snowfall versus rainfall in the western United States. *Journal of Climate, 19*, 4545–4559.

Lundquist, J. D., Neiman, P. J., Martner, B., White, A. B., Gottas, D. J., & Ralph, F. M. (2008). Rain versus snow in the Sierra Nevada, California: Comparing Doppler profiling radar and surface observations of melting level. *Journal of Hydrometeorology, 9*, 194–211.

Malamud-Roam, F. P., Ingram, B. L., Hughes, M., & Florsheim, J. L. (2006). Holocene paleoclimate records from a large California estuarine system and its watershed region: Linking watershed climate and bay conditions. *Quaternary Science Reviews, 25*, 1570–1598.

Metropolitan Water District of Southern California. (1931). *Rainfall and stream run-off in southern California since 1769*.

Minnich, R. A. (2007). *Climate, paleoclimate, and paleovegetation*. In M. Barbour (Ed.), *Terrestrial vegetation of California* (3rd ed., p. 43). University of California Press. http://doi.org/10.1525/california/9780520249554.003.0002

Mote, P. W., Hamlet, A. F., Clark, M. P., & Lettenmaier, D. P. (2005). Declining mountain snowpack in western North America. *Bulletin of the American Meteorological Society, 86*, 39–49.

Mount, J., & Dettinger, M. (2020, June). *California's twenty-first century*. Public Policy Institute of California (PPIC) blog.

NASA Solar Dynamics Laboratory. (2013). *NASA's SDO observes fast-growing sunspots*. https://www.nasa.gov/mission_pages/sdo/news/fastgrowing-sunspot.html

National Oceanic and Atmospheric Administration (NOAA). (2022). *National Centers for Environmental Information (NCEI) climate at a glance, statewide*. https://www.ncei.noaa.gov/access/monitoring/climate-at-a-glance

Public Policy Institute of California. (2020, June 23). *California's 21st century megadrought*. PPIC blog. https://www.ppic.org/blog/californias-21st-century-megadrought/

Purkey D. R., Joyce, B., Vicuna, S., Hanemann, M. W., Dale, L. L., Yates, D., & Dracup, J. A. (2008). Robust analysis of future climate change impacts on water for agriculture and other sectors: A case study in the Sacramento Valley. *Climatic Change, 87*, 109–122. http://doi.org/10.1007/s10584-007-9375-8

Roundtree, L. (1985). Drought during California's mission period 1769–1834. *Journal of California and Great Basin Anthropology, 7*(1), 7–20.

Roundtree, L., & Rayburn, R. (1980). Rainfall variability and California agriculture: An analysis from harvest and tree ring data. *Yearbook of the Association of Pacific Geographers, 42*, 31–44.

Sarkovich, M. (2009). Sacramento municipal utilities district's urban heat island mitigation efforts. In *2nd International Conference on Countermeasures to Urban Heat Islands*, September 21–23, 2009. https://heatisland.lbl.gov/sites/default/files/cuhi/docs/221120-sarkovich-ppt.pdf

Schimmelmann, A., Meixun, Z. C., Harvey, C., & Lange, C. B. (1998). A large California flood and correlative global climatic events 400 years ago. *Quaternary Research, 49*, 51–61. https://www.cambridge.org/core/journals/quaternary-research/article/abs/large-california-flood-and-correlative-global-climatic-events-400-years-ago/2BDC9161BF99A290926777D4746BBA77

Self, S., Zhao, J.-X., Holasek, R. E., Torres, R. C., & King, A. J. (1993). *The atmospheric impact of the 1991 Mount Pinatubo eruption*. USGS, cited by NTRS-NASA Tech. Reports Server. https://pubs.usgs.gov/pinatubo/self/

Warnert, J. E. (2014). *The California drought is helping return weather pattern to normal*. UC Division of Agriculture and Natural Resources: Green Blog, March 27, 2014.

Weare, B. C. (2009). How will changes in global climate influence California? *California Agriculture, 63*(2), 59–66. https://calag.ucanr.edu/archive/?article=ca.v063n02p59

Zelse, L., Garcia, Y., & Newsom, G. (2022). *Indicators of climate change in California* (4th ed.). California Environmental Protection Agency Office of Environmental Health Hazard Assessment. https://oehha.ca.gov/media/downloads/climate-change/document/2022caindicatorsreport.pdf

12

California Weather and Air Pollution

Fig. 12.1 San Francisco's skies are polluted from the September 9, 2020, wildfire smoke (*Credit* Aaron Maizlish/Flickr/Creative Commons BY 2.0)

Hello carbon monoxide, hello sulfur dioxide. The air, the air is everywhere.
—from Hair

San Francisco's unhealthy air due to wildfire smoke is seen in Fig. 12.1.

© The Author(s), under exclusive license to Springer Nature
Switzerland AG 2023
S. LaDochy and M. Witiw, *Fire and Rain*,
https://doi.org/10.1007/978-3-031-32273-0_12

Fig. 12.2 Los Angeles air in the 1950s and 60s brought tears to pedestrians' eyes (California Air Resources Board, 2019)

In parts of California, residents are becoming acquainted with the health effects of wildfire smoke. In recent years, wildfires, especially in northern California, have become more frequent and more destructive (see also Chap. 8: winds and wildfires). One study (Heft-Neal, 2020) found that wildfire smoke between August and September of 2020 led to thousands of excess deaths, just counting seniors (65 years and over). During the Dixie and Caldor fires of August 2021, the San Francisco Bay area had some of the highest Air Quality Index numbers in the world, rated extremely unhealthy (Figs. 12.1 and 12.2).

Smog and southern California seem to be paired historically with comments like breathing Los Angeles air is like smoking two packs of cigarettes a day. That may have been the case in the 1950s through early 1970s, but times and conditions have changed. Although Los Angeles still has the worst smog (ozone) of any United States city, much of the golden state is quite pristine in relation to pollution.

12.1   The American Lung Association 2022 "State of the Air"

The American Lung Association 2022 "State of the Air" report looked at three different categories of air pollution. Those categories include ozone, annual particle pollution, and short-term particles. Ozone pollution is often referred to as smog, while particle pollution includes aerosols, dust, and soot. When comparing United States cities, California cities led in all three lists. Los Angeles was worst in ozone pollution; Fresno was worst in annual particles and Bakersfield topped the short-term particle pollution list.

What makes Los Angeles and other California cities so polluted? Mostly, it has to do with population (and cars). But it also has to do with the Golden State's weather, climate, and topography. Lots of sun, years of low rainfall and more frequent wildfires in recent years, have caused the air quality index to register unhealthy for vulnerable groups. Los Angeles is notorious for having the worst smog in the country most years.

Among the cleanest counties in the nation for PM 2.5, were Lake, San Benito, and Nevada counties in that order. Among the cleanest counties in the United States for ozone were mainly northern California counties, with no unhealthy readings.

The ranking for the worst cities in the country for ozone and particulate air pollution are listed in Appendix D.

12.2 Ozone and Particulate Air Pollution

Ozone is a secondary pollutant formed by photochemical reactions between precursor nitrogen dioxide and organic hydrocarbons, or volatile organic carbons (VOC). These precursors mostly are a result of combustion in vehicles. Photochemical reactions are chemical reactions taking place in sunlight. The greater the sunshine and temperatures the greater is the ozone formation (An ozone molecule is composed of three oxygen atoms unlike our normal oxygen which has two. The extra oxygen can break off and cause damage to lungs).

Particulate matter (PM) is made up of both solid and liquid aerosols. Even particles too small to be seen can interfere with light, making the air hazy. Particles can be quite complex chemically, but size is important in how they affect us.

Particulates are classified according to size, from coarse to fine and ultra-fine. Coarse particles are measured as PM10, 10 µm in diameter or smaller. Fine particles, which are 2.5 µm (micrometers or microns) in diameter or smaller are called PM2.5. Ultrafine particles are smaller than 0.1 µm in diameter and can enter into the blood stream. All particles are harmful to your health, but the smallest are more dangerous to the body.

There are hundreds of different particle chemicals. The most common in cities are nitrates, sulfates, and black carbon. The last two are mostly from combustion, while nitrates are pollutants formed in chemical reactions and sunlight.

Particulate matter are practically everywhere, but are generally in higher amounts in urban areas. Cities have a lot of construction and road traffic that stirs up dust. In rural areas, agriculture and mining can be sources of particles. Fires add to the mix. Organics like spores and pollen can affect people with allergies. Chemical reactions in sunlight can even change gases into fine or ultrafine particles.

12.3   Why Is Air Pollution a Problem in California?

As mentioned above, California cities are some of the most polluted in the United States due to population, weather and climate, and the topography.

The estimated 40 million population (over twice the state's 1970 population) continues to grow. About 25% reside in Los Angeles County. In 2018, there were over eight million vehicles including 6.5 million cars in the county. Another nearly three million vehicles are in neighboring Orange County. The San Francisco Bay region and San Diego also add millions of vehicles. Cars remain the number one emitter of air pollution in the state.

Other transport modes contributing to the state's pollution include rail and ships. Both rail and shipping contribute large amounts of pollutants to the air. The Los Angeles-Long Beach port is the largest United States ship terminal. Arriving ships burning marine crude oil has been a health concern to neighboring communities. Gas-burning trucks unload ship containers and load nearby railroads creating more noise and pollution.

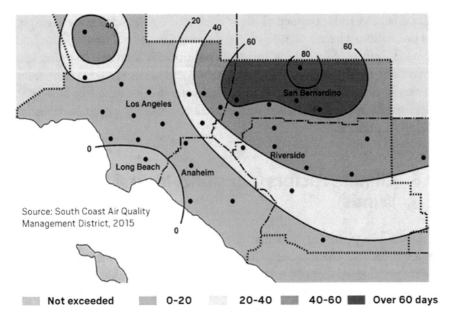

Not exceeded **0-20** **20-40** **40-60** **Over 60 days**

Fig. 12.3 Number of days exceeding the PM2.5 state standard in SW California during 2015 (South Coast Air Quality Management District, 2019). The highest values are in purple along the foothills of the San Bernardino Mountains, while coastal areas are upwind of the cities' pollution sources

12.4 Weather and Topography

The ocean breeze plays an important role in blowing pollutants downwind. Local mountains can either block the breeze or funnel it through passes (see Chap. 8). As most Californians live along the coast, their cars and industries present the largest source of pollution. The breeze in late morning and early afternoon pushes these emissions inland where they concentrate. In the Los Angeles Basin, highest ozone and particulate levels are found in inland valleys and along mountain slopes downwind of cities (Fig. 12.3).

12.5 Terrain: A Perfect Place for Smog

California's surface features (its topography) coupled with its climate make it ideal for the formation and persistence of pollution. During summer, ozone forms in the strong sunlight acting on pollutants from transportation. This is particularly true in the cities of Los Angeles and San Jose. With mountains or hills encircling the cities in all directions except on the coastline, pollutants tend to be trapped. This is enhanced by the presence of temperature inversions, where a warm layer aloft prevents the pollutants from being dispersed upward. Since the solution to pollution is dilution, trapping smog horizontally by the surrounding mountains, and vertically with temperature inversions, dispersion is limited in coastal valleys. The urban buildings also contribute by reducing winds through friction, enhancing stagnation and the buildup of pollutants.

12.6 Climate Factors (Sunshine, Inversions, Winds)

California has plenty of sunshine (see Chap. 3). Sunshine produces higher temperatures and chemical reactions among pollutants. Nitrogen dioxide, a reactive gas formed from burning fuel in sunlight, has a reddish-brown color, which gives smog its characteristic brown appearance, reducing visibility. On smoggy summer days, the brown haze sometimes hides the picturesque mountains next door.

On hot, sunny days, nitrogen oxides and volatile organic compounds mostly emitted in vehicle combustion react in sunlight forming ozone. Smog is commonly identified by the presence of ozone. Rain, which cleanses the air, is at a minimum during the summer and early fall. The smog season, when ozone and particulates are highest, occurs during these dry months. Droughts and drier than normal years generally have much less days of rain and higher air pollution. Fall fires also add to diminishing air quality.

Wind, especially in summer, is quite low over much of the state, especially in the south. At lower latitudes, the Pacific High is more dominant much of the year with typically clear skies and weak winds. It is only during the winter and early spring, that storms and more cloudiness accompany cooler temperatures decreasing air pollution levels. Along the coast, the number of storms and days of rain decrease from north to south.

Temperature inversions occur throughout California, especially in summer. In general, the stronger the inversion, which is the difference of temperature between the warmer top and the cooler bottom of the inversion layer, the higher the air pollution. Strong inversions happen when there is a high-pressure system over California. Inversions are not always present with weak high-pressure or low-pressure systems and when they do occur are not very strong (Figs. 12.4 and 12.5). Inversion frequency also varies from year to

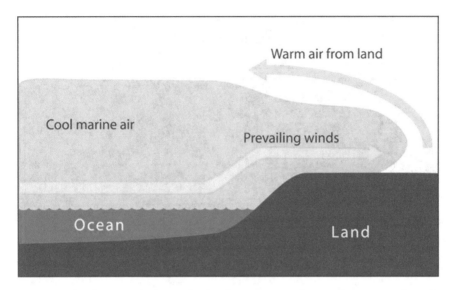

Fig. 12.4 When warm air from the land overlays cooler marine air, an inversion forms (Huber-California State University Northridge Geog103 notes)

year, partly following changes in sea surface temperatures along the California coast. Coastal warming, such as El Niño events, leads to weaker and less frequent inversions. Recent ocean warming also has impacted inversions in a similar way.

12.7   History of California Air Quality

The first significant smog episode occurred in Los Angeles in the early 1940s. The smog was so bad, that you couldn't see more than three blocks. People suffered from "pink eye", watery eyes, and difficulty in breathing. It was originally blamed on a nearby chemical plant. But smog was so pervasive that in 1947, the Los Angeles County Air Pollution Control District (APCD) was formed. It was the first air pollution agency in the nation. The Los Angeles County APCD regulated some industries, but it was cars that were the main cause.

The cause of ozone was discovered by Dr. Haagen-Smit (Fig. 12.6). It was the volatile organic compounds in gasoline, and oxides of nitrogen (NO_x) produced by the high temperatures in vehicle engines that reacted in sunlight to produce ozone. It was the ozone that caused the eye irritation and respiratory problems affecting Los Angeles residents (see Fig. 12.2).

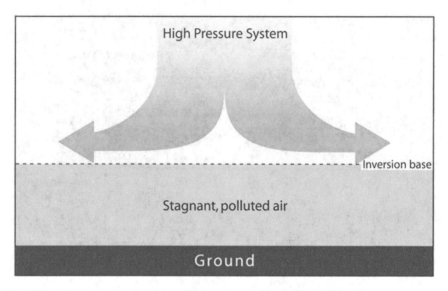

Fig. 12.5 In summer, a dominant high pressure along the California coast causes a subsidence inversion, with sinking warm air overlaying the cooler marine air (Huber-California State University Northridge Geog103 notes)

Fig. 12.6 Dr. Haagen-Smit, Cal Tech chemist, collects air samples in downtown Los Angeles in the 1950s (California Air Resources Board, 2019)

In 1950, the Los Angeles County APCD District banned smudge pots used to keep citrus orchards from freezing. Wind machines replaced smudge pots. Also banned was residential trash burning in 1957. Before then, most people burned their rubbish using small backyard incinerators.

In 1975, California led the way requiring all new cars sold in the state to have catalytic converters, with the United States and other countries adopting similar laws. In 1984, The California Smog Check program was authorized of all licensed vehicles. The program only passed clean running cars. Ozone levels in California began dropping beginning in the 1970s. Even as the car population doubled since then, the cleaner autos and reduced emissions from industries have led to bluer skies.

In 1996, the nation's automakers vowed to make zero-emissions vehicles, with General Motors introducing the EV-1. In 2003, car makers sued the state to eliminate any zero-emission vehicle standards. However, California again became a leader by banning the sale of new gas engine cars in the state by 2035.

12.8   Air Pollution Forecasting

Forecasting the levels of air pollution depends on the number of emissions and the dispersion horizontally and vertically, unless there is precipitation. Assuming the daily emissions to be fairly uniform, except slight decrease in car travel on weekends, then forecasted levels would depend mostly on dilution factors. Those factors include estimated winds and the vertical mixing height. The latter varies with the weather patterns, being higher (better dilution) when there is little or no inversion layer (warmer air overlying cooler air, Fig. 12.4). Lower, stronger inversions limit vertical mixing and can lead to elevated air pollution. This occurs when the Pacific high pressure is nearby. The sinking air with high pressure compresses and warms the upper air causing a subsidence inversion (sinking air as shown in Fig. 12.5). In coastal southern California, the average inversion height is about 2000 ft in summer when the Pacific High is closest to coastal California, while in winter the High is weaker and farther south leading to higher and weaker inversions.

Temperatures also are included in estimating pollution levels since photochemical smog increases with temperature. Photochemical reactions between different air pollutants speed up at higher temperatures, thus also favoring summer as the worst pollution season. Seasonal average ozone levels, (Fig. 12.7) show that ozone peaks in southern California in late summer when temperatures are highest. Smog correlates highly with the 850 mb (upper air pressure level) temperatures.

Wind direction as well as wind speed can influence geographic distribution of pollutants. Northeast Santa Ana winds may increase temperatures in southern California, but may also sweep pollutants out to sea.

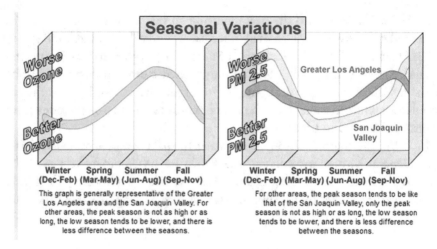

Fig. 12.7 Seasonal variations of ozone, PM in metro Los Angeles and San Joaquin Valley (California Air Resources Board, 2022)

12.9   Seasonal Variations

The main air pollutants associated with smog in California are ozone and particulates (PM10 and PM2.5). Ozone is a secondary pollutant formed by photochemical reactions between precursor nitrogen dioxide and organic hydrocarbons or VOCs (volatile organic carbons). Photochemical reactions are chemical reactions taking place in sunlight. The greater the sunshine and temperatures the greater is the ozone formation. Therefore in most of California, the peak season for ozone is mid-summer to early fall when temperatures are highest and high pressure leads to more sunlight. Winter is the rainy season along with reduced solar input, so that would typically lead to lowest pollution levels. Winter rain, and especially snow, removes pollutants. So for at least one day, the air may remain relatively clean. Even without rain, clouds reduce heating and photochemical reactions.

12.10   Diurnal Variations

On hot, summer days, urban runners are advised to exercise either in early morning or evening hours. Although the morning rush hours produce more pollutants, the temperatures are cool and photochemical reactions are weak. Nitrogen oxide peaks with traffic, and then as the day warms up, photochemical reactions convert NO to NO_2, nitrogen dioxide, which reacts in sunlight with VOCs producing ozone and other secondary pollutants. Ozone peaks

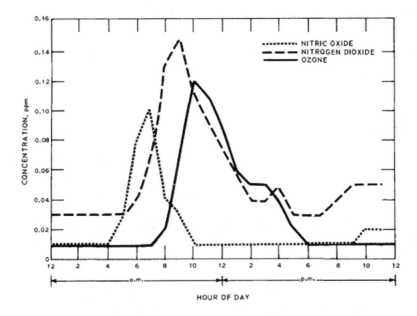

Fig. 12.8 Diurnal variations of NO, NO₂, and ozone concentrations in Los Angeles on July 19, 1965 (National Air Pollution Control Administration, United States Department of Health, Education, and Welfare)

in the afternoon when temperatures are highest, though along the coast sea breezes may shift smog further inland. By late afternoon, with cooling and less direct sunlight, ozone decreases. A secondary NO_x peak may occur in the late afternoon rush traffic, but with less sunlight, ozone is not as high as in early pm. Far inland, the smog front may reach inland cities and up into the foothills later in the day. Figure 12.8 shows the typical summer diurnal pattern of air pollutants in southern California.

12.11 Future California Air Pollution

While the state has made great strides in reducing air pollution emissions, the obstacles to further improvements are formidable. The population of the state continues to increase as does the number of vehicles. Temperatures continue to increase regionally and locally (urban heat islands). Smoke from fires has increased lately. While ozone levels are declining with cleaner cars, particulate air pollution has not improved in recent decades. Significant improvement in air quality did occur due to the recent COVID-19 lockdown. Californians saw how less traffic changed air quality and urban congestion for the better. If the state has the will to solve its transportation problems, say with electric or hydrogen vehicles, the air may become cleaner than before.

Bibliography

California Air Resources Board. (2019). *History*. CA Air Resources. https://ww2.arb.ca.gov/about/history

California Air Resources Board. (2022). *Air quality and emission data*. https://ww2.arb.ca.gov/CAPP-air-quality

Childs, J. W. (2019, April 24). U.S. cities with the worst air pollution; The list is out. *The Weather Channel*. https://weather.com/news/news/2019-04-24-most-air-polluted-cities-us-2019-climate-change

Heft-Neal, S. (2020). Indirect mortality from recent wildfires in CA. *G-FEED Post*. http://www.g-feed.com/2020/09/indirect-mortality-from-recent.html

Huber, A. (2004). *Physical geography notes. Smog and inversions*. http://www.csun.edu/~hmc60533/CSUN_103/weather_exercises/soundings/smog_and_inversions/.htm

National Air Pollution Control Administration, USDHEW. (1969). *Air quality criteria for photochemical oxidants*. AP-69. https://nepis.epa.gov/

Rice, D. (2019, April 24). Bad air days are on the rise: The nation's most polluted city is… *USA Today*. https://www.usatoday.com/story/news/nation/2019/04/24/air-pollution-smog-soot-worst-california/3551734002/

Singh, M., Phuleria, H. C., Bowers, K., & Sioutas, C. (2006). Seasonal and spatial trends in particle number concentrations and size distributions at the children's health study sites in Southern California. *Journal of Exposure Science & Environmental Epidemiology, 16*(1), 3–18. https://pubmed.ncbi.nlm.nih.gov/16077742/

South Coast Air Quality Management District. (2019). *2019 annual report*. SCAQMD. https://www.aqmd.gov/docs/annual-reports/2019-annual-report.pdf

The Daily Breeze. (2022, December 16). *Investigation finds LA Harbor-area smog challenges grow as new health threats emerge*. https://www.dailybreeze.com/2018/08/07/investigation-finds-la-harbor-area-smog-challenges-grow-as-new-health-threats-emerge/

13

California Agriculture

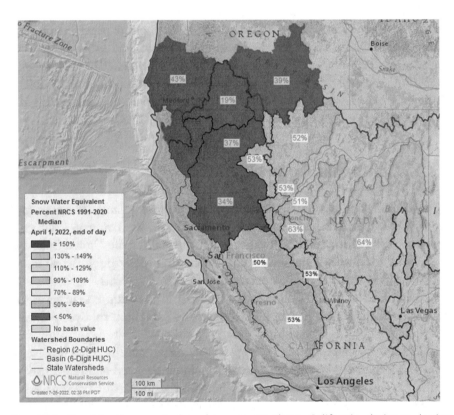

Fig. 13.1 Possibly the worst drought ever. Northern California drainage basins shown in red were less than half the normal snow water equivalent (United States Department of Agriculture: Natural Resources Conservation Services)

© The Author(s), under exclusive license to Springer Nature
Switzerland AG 2023
S. LaDochy and M. Witiw, *Fire and Rain*,
https://doi.org/10.1007/978-3-031-32273-0_13

California is a garden of Eden, a paradise to live in or see
But believe it or not, you won't find it so hot, if you ain't got the do, re, mi.
—Woody Guthrie (Do-Re-Mi)

The year 2022 was the third dry year in a row for California, making this one of the worst droughts for the state that has seen so many droughts. While the 2012–2016 drought was thought to be the worst in the last 1000 years, the 2020–22 drought may be even worse due to the record heat. In fact, the whole twenty-first century has had below normal precipitation. Snowpack in the Sierra Nevada has also been subpar with hardly any in the 2021–22 snow season by early spring. The Colorado River Basin wasn't much help. State reservoirs in 2022 suffered low levels as well. Farmers had little choice with rationed surface water to dig deeper wells. In the previous 2012–16 drought, over 2000 wells went dry, with agricultural losses in the billions of dollars along with thousands of job losses. With another severe drought hitting the state, the future of farming may be in jeopardy (Fig. 13.1).

13.1   California Agro-business

Agriculture includes animal husbandry, horticulture (plants, flowers), and crops. California is the leading agricultural state in the country. California produces 13% of the total cash value in agricultural for the United States. It is the lone producer of numerous crops including many nuts like almonds and walnuts and fruits such as nectarines, dates and apricots. Agriculture in California is about a $50 billion industry—first in the nation. Dairy was the leading agricultural product in 2021 valued at $7.0 billion that year. Grapes follow at $5.2 billion, then almonds at $5.0 billion.

Because of its many microclimates, California agriculture must be diverse in growing methods across the state that includes irrigated and dry farming. California's drought has been especially hard on the production of crops, especially orchards. Nearly half of California's farms use irrigation, with crops like almonds being especially thirsty.

With its mild, Mediterranean climate, California produces more than 400 agricultural products. Much of the nation's fruits and nuts and over a large portion of its winter vegetables are grown on state farms.

California agriculture dominates much of the arable land of the state, using most of its water, both above and below ground, much of its land, and changes its environment, as well as its weather and climate. We now look at the crop-weather relationships.

13.2 Grapes and Climate

In 1997, the Robert Mondavi Chardonnay wine, in a blind sampling in the Grand European Wine Tasting, was judged number one by mostly French judges. This amazed the wine world and put California wines on the map. The year before a California Cabernet Sauvignon won the French wine tasting award for reds. California wine and table grapes have a long history.

In 1839, the first table vineyard was established in the small Mexican town now known as Los Angeles. Grapes from here were shipped to northern California and later to the east on the newly built railways.

Today, the grape industry is a multi-billion-dollar enterprise, with over 80 different varieties grown during the summer in red, green, and black colors (Fig. 13.2). Table grapes and wines are transported globally.

Grapes, especially wine grapes, have expanded among the state's coastal valleys and Central Valley. If it were a nation, California would be the fourth largest producer in the world, trailing only France, Italy, and Spain. There are more than 139 *American Viticultural Areas* (AVAs) in California (distinct wine grape growing areas recognized by the US government), resulting from the diverse microclimates of the state.

So what makes California more suitable for growing a wide variety of wine grapes compared to other states? One of the leading factors is the Mediterranean climate with cool Pacific Ocean breezes near the wine-growing regions. The cool sea breeze helps modify the warm, dry summer air inland. Summer temperatures during the growing season are warm, but a few days during the summer can exceed 100 °F. These growing conditions of mild, wet winters and warm, dry summers under abundant sunshine allow for grapes to mature and the sugar to turn into alcohol. The sugar content is controlled by sunshine, temperature, and rainfall. In warmer climates, the grapes can mature faster, producing higher sugar levels and a darker color, and higher alcohol levels. In cooler climates, the sugar content is much less, tasting tart, and a lower alcohol level. Bottoms up?

Fig. 13.2 Despite warming, fires, and droughts, California grapes continue to be one of the global leaders. Purple and green grapes are displayed at grocer (Photo by S. LaDochy)

13.3 Crops

Owing to its mild climate, California can grow hundreds of different crops, many can only be grown in the Golden state. The attractiveness of mild, subtropical climate and orange trees in the backyard lured many Midwesterners to the state. John Steinbeck's *Grapes of Wrath* also described the state as an agricultural Eden growing winter crops that were not grown in other parts of the country.

California's other advantage is its long growing season which allows it to grow a variety of products. California leads the nation in many of these crops, including lettuce and tomatoes, its most important cash crops.

13.3.1 Rice

Rice is a water-intensive crop that requires fields to be flooded. Rice can be grown on less suitable land for most other crops. Most rice is grown in the Sacramento Valley. Here days are sunny and warm, while nights are quite

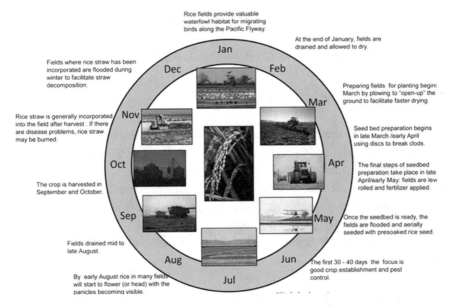

Rice fields provide valuable waterfowl habitat for migrating birds along the Pacific Flyway.

At the end of January, fields are drained and allowed to dry.

Fields where rice straw has been incorporated are flooded during winter to facilitate straw decomposition.

Preparing fields for planting begins March by plowing to "open-up" the ground to facilitate faster drying.

Rice straw is generally incorporated into the field after harvest. If there are disease problems, rice straw may be burned.

Seed bed preparation begins in late March / early April using discs to break clods.

The final steps of seedbed preparation take place in late April / early May: fields are level rolled and fertilizer applied.

The crop is harvested in September and October.

Once the seedbed is ready, the fields are flooded and aerially seeded with presoaked rice seed.

Fields drained mid to late August.

The first 30 - 40 days the focus is good crop establishment and pest control.

By early August rice in many fields will start to flower (or head) with the panicles becoming visible.

Fig. 13.3 Summary of the annual growing cycle for rice in California (University of California, Davis, 2022)

cool. The clay soils here also retain the water. Much of this rice is exported. Drought and extreme heat both reduce rice yields.

In terms of climate change, rice production releases substantial amounts of agricultural greenhouse gas emissions, mainly methane, a much more potent greenhouse gas than carbon dioxide. Climate warming may lead to more greenhouse gases released from rice fields. However, new techniques in organic farming can reduce water use and released methane (Fig. 13.3).

13.3.2 Alfalfa

Alfalfa (hay) is the main feeding stock for livestock. Since dairy is the state's leading agricultural product, alfalfa is the largest crop in area and makes California the leading hay-producing state in the nation. It also uses up the most water of any California crop. With California's year-long growing season, alfalfa can be grown all year-round. California easily outproduces the rest of the country. Most irrigated alfalfa is produced in the Central Valley, particularly San Joaquin Valley. The Southeast Desert also grows lots with even higher yields than other regions. This region can harvest year-around with twice the yields of the rest of the state. Water is the limiting factor on yields. Alfalfa requires huge amounts in the Central Valley and even more in the Coachella Valley, which presents problems in drought years.

13.3.3 Almonds

The California almond industry is worth billions annually. California grows 100% of almonds in North America and 80% of almonds globally. But almonds require a lot of water. During droughts almonds have lower yields, while pistachios are less thirsty. Pistachio trees are much better suited to drier climates than almonds and even have a longer life span. It takes about a gallon of water to grow one almond, and several gallons to produce a walnut. Almonds use more water than most of southern California cities (Figs. 13.4 and 13.5).

Fig. 13.4 Almonds are very thirsty exported nuts using lots of California's precious water (Photo by S. LaDochy)

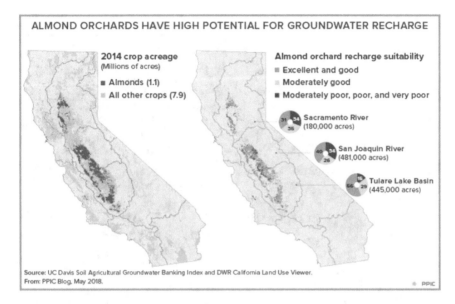

Fig. 13.5 Droughts impact almond-growing areas in Central Valley often leading to more groundwater withdrawals (Hanak, 2018)

13.4 California Water Transfer System

California's agricultural success requires irrigation. The California Department of Water Resources estimates that in an average year, some 9.6 million acres are irrigated with about 34 million acre-feet of water. Most irrigation water is used efficiently, though not always. When precipitation does not meet evapotranspiration (ET) losses, irrigation is needed for crops. Much of this irrigation is provided with one of the remarkable achievements in modern agriculture, the State Water Project.

The California State Water Project (SWP) is a water system of reservoirs and aqueducts carrying water from the more abundant north to the drier south, traveling roughly across two-thirds the length of California (Fig. 13.6). SWP water irrigates vast acreage of farmlands, mainly in the San Joaquin Valley, but also services cities and industries in coastal central and southern California. The other main sources of water are the Colorado River and the LA Aqueduct (see Chap. 2).

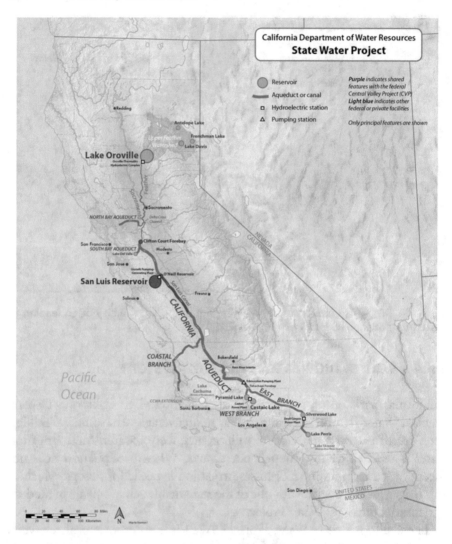

Fig. 13.6 State Water Project (California Department of Water Resources)

13.5 Dairy

One cow can produce about 6.3 gallons of milk a day and several thousands of gallons of milk in a lifetime. California is the leading dairy state in the United States; however, cow manure is also the largest source of California's methane emissions. It's estimated that California's nearly two million dairy cows produce over half of the state's methane emissions. With new technologies, California dairy farms could reduce methane substantially.

California's milk production has grown tremendously in recent decades, to feed America's growing population and its need for milk. California's mild climate favors dairy over other states as cows do not require shelter and feed (alfalfa) is grown year-round.

Much of California's dairy farms are in the Central Valley, where it's warm for over five months of the year. At very high temperatures, cows feel heat stress. With stress, cows produce less milk. Long heat spells increase heat stress in dairy cows. During the July 2006 heat wave record temperatures (Fresno hit 112 °F), tens of thousands of cows died from the heat. California is also the second leading producer (state) of cheese, with nearly half of all milk going into making numerous varieties of cheese products. And don't forget butter.

13.6 Horticulture

The flowers and house plants industry also depend on stable weather conditions, not extremes. Too hot, too cold, too dry, or too wet can lead to poor or no yields. For spring-flowering perennials, milder temperatures can stretch out their flowering, while too much heat can kill plants. California's mild seasons provide year-round growing conditions for a variety of plants and flowers.

Even mid-range temperatures matter, and not just to pollinators like honeybees, which start working at about 55 °F. Tomatoes need warm night-time temperatures above 55 °F. Sensitive plants often have a narrow range of tolerance to temperatures, sunlight, wind, and water. Climate change may favor some plants while diminish others. Gardeners can estimate suitability for growing sensitive plants by average climatic information, such as minimum temperatures, used for a hardiness zones map (Fig. 13.7). However, the possibility of extreme weather should also be considered.

California is the leading producer of cut flowers and garden and house plants in the United States, with over a billion dollars sold in 2018 in flowers alone. Gardening has grown as a thriving business for urban gardeners. Heat and drought are big concerns, although growers can switch to more heat-tolerant flowers, such as sunflowers and marigolds, or drought-tolerant plants. To replace thirsty lawns in southern California, many turn to succulents and cacti. The largest, yet unknown amounts of revenue come from selling legal or illegal marijuana crops.

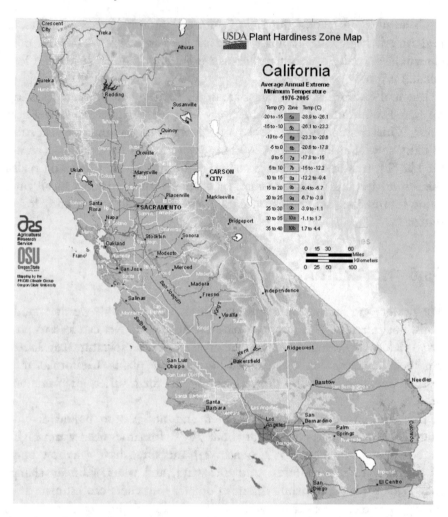

Fig. 13.7 Plant hardiness zone map helps gardeners estimate the thermal conditions in various zones. As minimum temperatures are important to sensitive plants, they are used for the hardiness zones (United States Department of Agriculture, 2022)

13.7   Hazards to Agriculture

In California, minimum temperatures are rising faster than maxima. That is good for winter crops, but not so good for orchards that require "chill" hours (hours with temperatures less than 50 °F).

13.7.1 Frosts and Freezes

Even in milder coastal valleys, the frost is a big concern of farmers. Figures 13.8 and 13.9 show the mean dates for the first and last killing frost (28 °F) based on weather records from 1980 to 2010. No frost is shown for the southernmost California, much of coastal plains and the Sacramento Valley area receiving Bay's marine air. However, these areas have suffered from frost in extreme years, especially before 1980. Since temperatures generally fall with elevation, mountainous regions are less suitable for most agriculture due to shorter growing seasons and early frost.

Frost prevention techniques in California include heaters, wind machines, watering, mulching, and coverings. Smudge pots, kerosene heaters, were used to keep southern California orchards from freezing by not only direct heating but also trapping heat with a layer of polluted smoke. Air pollution regulations eventually outlawed the practice. Wind machine, with or without

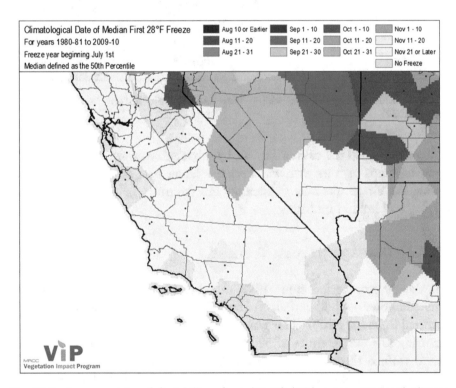

Fig. 13.8 Average date of first killing frost (28 °F) (Midwestern Regional Climate Center, 2022)

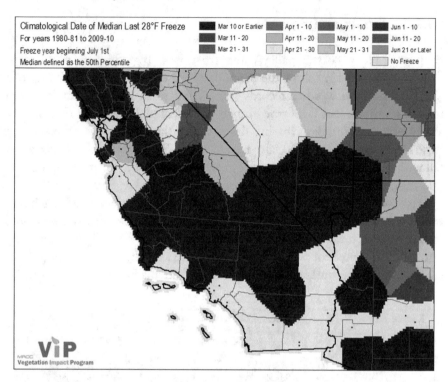

Fig. 13.9 Median day for last killing frost (28 °F) (Midwestern Regional Climate Center, 2022)

heaters, breaks up the cold surface air by mixing downward warmer air aloft. Spraying crops or trees with water releases latent heat as water turns to ice, raising the air temperature. Mulching and coverings help reduce radiation cooling and hold in moisture.

13.7.2 Damaging Frosts/Freezes

For the 30-year period from 1992 through 2021, extreme cold weather resulted in over 4.4 billion dollars of crop damage. However, five of these days occurred in January 2007. This was an exceptionally cold month with Fresno reporting low temperatures of 32 °F or below for 19 consecutive days. The period from 6 through 14 January 2007 was especially damaging. During this period, crop damage occurred from the San Francisco Bay area to the Coachella Valley.

On January 6, record cold resulted in over $20 million in damage from Marin County in the San Francisco Bay area to the Monterey Bay area. Temperatures in some places fell as low as the upper teens.

After the $20 million loss on January 6, damaging frosts and freezes returned and occurred on every day from January 11 through January 14. The amount of crop damage was incredible. In the San Joaquin Valley alone, losses exceeded $700 million. Total damage for the state was $1.3 billion for this four-day period.

Other months during the 30-year period that recorded over $30 million in crop damage were December 1998 and April, 2001 where a late season frost resulted in damage to grapes and other fruit in the San Joaquin Valley.

13.7.3 Floods, Droughts, and Heatwaves

Floods and droughts impact agricultural operations and yields. The recent drought in 2012–2016 had direct costs to agriculture totaling several billion dollars. Losses in farm employment were over 17,000 jobs. Water withdrawals in wells caused parts of the Central Valley to sink as much as two feet per year (Fig. 13.10) even though pumping became more expensive. Heat waves during droughts increase this impact. Floods on the other hand can be even more costly. Flooded lands are deadly to livestock. In 1862, the storms drowned an estimated 200,000 cattle. Today, there are millions more cattle including dairy cows in the Central Valley (which was flooded in 1862). Crops in the Central Valley represent billions of dollars in revenue to the state and can be severely damaged by floods. Floods and droughts are discussed in greater detail in Chap. 6.

13.8 Climate Change

Although many reports imply climate change (Chap. 11), particularly warming temperatures, will have an ill effect on agriculture, some aspects of warming are advantageous. For instance, in the last 50 years, frost days, the biggest hazard to vineyards, has decreased by about two weeks, leading to a greater abundance of grapes and higher quality. Droughts and heat waves, on the other hand, have caused great harm to both crops and livestock. Fruit trees also require a winter chill, which are less frequent with warming. Droughts, which have been longer and more frequent in the twenty-first century, have strained state water supplies. Irrigation uses some 80% of the available freshwater in the state. Without the elaborate water transport from

Figure 3.17: Cumulative Change in Statewide Groundwater Levels, 2012-2016 Drought

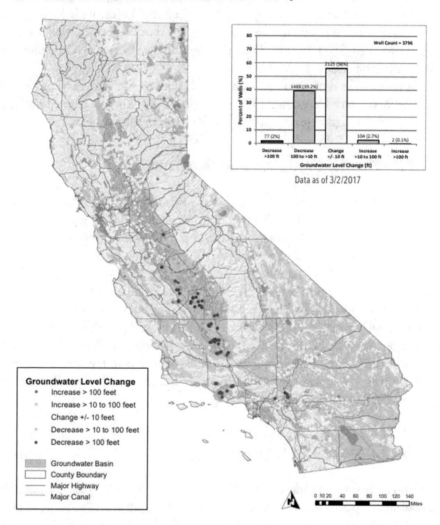

Fig. 13.10 Groundwater losses from the 2012–2016 drought shows the San Joaquin Valley is a huge tapper of groundwater (California Department of Water Resources)

the wetter northern part of the state to the drier south, large-scale agriculture would not be possible. Some portions of the San Joaquin Valley use the equivalent of 60 in. annually of man-made rain (irrigation) to support crops and orchards.

Trees present a unique problem with warming since the trees can't move. A fruit or nut tree planted now may not be suited to climatic conditions by the time it begins bearing fruit in a decade or two. Farmers need to know what to plant today to match possible climates of the future.

Central Valley temperatures are predicted to rise by several degrees Fahrenheit by 2100, with extreme heat events expected to lengthen. Winters are beginning to be too warm to permit trees important "chill" hours needed for flowering (see Chap. 11).

Irrigation water is becoming saltier too. During droughts deeper wells often hit saltier water. A new state law, the Sustainable Groundwater Management Act, started in 2020, regulates how much water farmers can withdraw from the Central Valley aquifer. The law promises to bring more equable change to Valley agriculture. Irrigation water depends mostly on abundant winter snow in the Sierras and spring melted runoff from snow for the dry summers. That snowfall is becoming less reliable.

The Central Valley's famous Tule fog, needed to lower evaporation rates and maintain crucial chill hours, is also declining. Particulates in the Central Valley, which attract moisture droplets and help create the fog, are becoming less. Since the 1970 Clean Air Act, pollution has declined. While that helps air quality, it does not help agriculture that has come to depend on fog.

Climate change can also affect agriculture with its effects on pests and diseases. More extremes in temperatures and precipitation, including less snow and earlier snow melt may impact several growing areas. Heavy rains do not help either. Fruit trees, like cherries, can split if the fruit takes up too much water at its mature stage, and strawberries are susceptible to mold if they are water-logged. Another big concern has been hail that could damage fruit; hail may be more frequent with climate change. Larger and more severe wildfires have invaded agricultural land.

On the positive side, increases in carbon dioxide should stimulate plant growth leading to higher yields of many crops. But there is a limit to that as well.

Bibliography

California Dairy Pressroom & Resources. (2022). *Real California dairy facts*. Real California Milk. https://www.californiadairypressroom.com/Press_Kit/Dairy_Industry_Facts. Accessed December 12.

California Department of Food and Agriculture. (2009, July). *Climate change and agriculture*. https://www.cdfa.ca.gov/agvision/docs/Climate_Change_and_Agriculture.pdf

California Department of Food and Agriculture. (2022). *California agricultural production statistics.* https://www.cdfa.ca.gov/statistics/. Accessed September 1.

California Department of Water Resources. (2022a). *Agricultural water efficiency.* https://water.ca.gov/Programs/Water-Use-And-Efficiency/Agricultural-Water-Use-Efficiency. Accessed October 11.

California Department of Water Resources. (2022b). *State water project.* https://water.ca.gov/Programs/State-Water-Project. Accessed December 16.

California Tablegrapes Commission. (2022). *All about grapes.* https://www.grapesfro mcalifornia.com/all-about-grapes/. Accessed December 13.

CBS News. (2016, December 16). *Waste from California dairy farms presents climate change challenge.* https://www.cbsnews.com/sanfrancisco/news/waste-from-califo rnia-dairy-farms-presents-climate-change-challenge/

Famiglietti, J., Lo, M., Ho, S. L., Bethune J., Anderson, K. J., Syed, T. H., Swenson, S. C., de Linage, C. R., & Rodell, M. (2011). Satellites measure recent rates of groundwater depletion in California's Central Valley. *Geophysical Research Letters, 38*(3). https://doi.org/10.1029/2010GL046442

Graff, Z. (2019, June 3). *How weather impacts wine grapes.* Weatherworks. https://weatherworksinc.com/news/climate-wine-grapes

Hanak, E. (2018). Three water challenges for almonds. PPIC blog, May 31, 2018.

Hooker, B (2013, January 7). *California rice is ahead of the curve on reducing greenhouse gases.* NewGenFarmer. https://www.farmprogress.com/rice/california-rice-ahead-curve-reducing-greenhouse-gases

Inside Climate News. (2020). *Cows get hot too: A new way to cool dairy cattle in California's increasing heat.* https://insideclimatenews.org/news/04122020/califo rnia-dairy-agriculture-cows-cattle-climate-change-drought/

Klemann, D. (2020, March 26). *California's flower industry gutted from coronavirus as local farms wilt under financial losses.* KSBY. https://www.ksby.com/news/local-news/californias-flower-industry-gutted-from-coronavirus-as-local-farms-wilt-under-financial-losses#:~:text=California%20is%20the%20bouquet%20of%20the%20flower%20industry,Waldo%20Emerson%20wrote%20the%20world%20laughs%20in%20flowers

Lund, J., Medellin-Azuara, J., Durand, J., & Stone, K. (2018). Lessons from California's 2012–2016 drought. *ASCE Library, 4*(10). https://doi.org/10.1061/%28ASCE%29WR.1943-5452.0000984

Lurie, J. (2015, January 12). California's almonds suck as much water annually as Los Angeles uses in three years. *Mother Jones.* https://www.motherjones.com/env ironment/2015/01/almonds-nuts-crazy-stats-charts/

Midwestern Regional Climate Center. (2022). *Freeze date tool.* https://mrcc.purdue.edu/freeze/freezedatetool.html

Nemani, R., White, M., Cayan, D. R., Jones, G., Running, S., Coughlan, J., & Peterson, D. (2001). Asymmetric warming over coastal California and its impact on the premium wine industry. *Climate Research, 19*, 25–34.

Putnam, D. H., Sommers, C. G., & Orloff, S. B. (2007, December). *Alfalfa production systems in California.* https://alfalfa.ucdavis.edu/IrrigatedAlfalfa/pdfs/UCAlfalfa8287ProdSystems_free.pdf

Schapiro, M. (2019, July 11). *California's Central Valley is designing the future of American agriculture.* Pacific Standard. https://psmag.com/environment/farmers-are-salty-over-soils-saline-levels

Silver Lakes Farm. (2022, December 16). *How much milk does a cow produce in a day?* https://silverlakefarms.com/how-much-milk-does-a-cow-produce-in-a-day/

Stokstad, E. (2020, April 16). Deep deficit. *Science.* http://doi.org/10.1126/science.abc2671. https://www.science.org/content/article/droughts-exposed-california-s-thirst-groundwater-now-state-hopes-refill-its-aquifers

Stroop, R. (2017, June 26). *California's top 10 ag products.* Farm Flavor. https://farmflavor.com/california/californias-top-10-ag-products/

United States Department of Agriculture. (2022a). *USDA plant hardiness zone map.* https://planthardiness.ars.usda.gov/pages/map-downloads. Accessed December 16.

United States Department of Agriculture. (2022b). *Snow water equivalent map.* USDA Natural Resources Conservation Service. https://www.nrcs.usda.gov/wps/portal/wcc/home/quicklinks/imap. Accessed December 16.

University of California, Davis. (2022). *About California rice.* https://rice.ucanr.edu/About_California_Rice/. Accessed December 16.

Wein, A., Mitchell, D., Peters, J., & Rowden, J. (2016). Agricultural damages and losses from ARKstorm scenario flooding in California. *Natural Hazards Review, 17*(4). https://doi.org/10.1061/%28ASCE%29NH.1527-6996.0000174

Wikipedia. (2020, April 27). *Grand European jury wine tasting of 1997.* https://en.wikipedia.org/wiki/Grand_European_Jury_Wine_Tasting_of_1997

Wine Institute. (2022, November). *Wine statistics by the numbers.* https://wineinstitute.org/our-industry/statistics/

Appendix A: 1991–2020 Normals

Source: https://scacis.rcc-acis.org/.

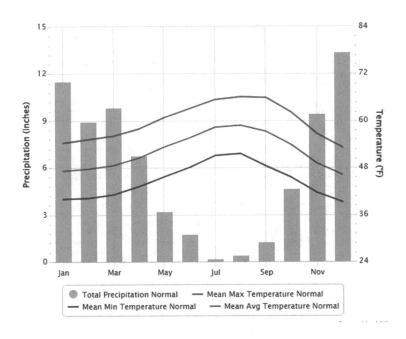

Monthly Climate Normals (1991–2020) – CRESCENT CITY 3 NNW, CA

Month	Total precipitation normal (in.)	Mean max temperature normal (°F)	Mean min temperature normal (°F)	Mean avg temperature normal (°F)
January	9.4	56.2	41.9	49
February	6.94	56.2	42.4	49.3
March	7.81	55.8	42.8	49.3
April	5.12	57.2	44.4	50.8
May	2.48	59.3	47.2	53.3
June	1.4	61.6	49.3	55.5
July	0.33	63	51.6	57.3
August	0.37	64.2	52.8	58.5
September	1.1	64.5	50.2	57.3
October	4.07	62.3	47.2	54.8
November	7.82	58.2	44	51.1
December	11.14	55.2	41.4	48.3
Annual	57.98	59.5	46.3	52.9

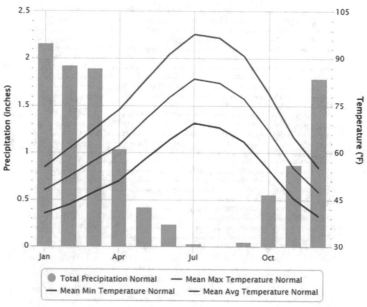

Monthly Climate Normals (1991–2020) – FRESNO YOSEMITE INT'L, CA

Powered by ACIS

Month	Total precipitation normal (in.)	Mean max temperature normal (°F)	Mean min temperature normal (°F)	Mean avg temperature normal (°F)
January	2.16	55.4	40.6	48
February	1.93	61.3	43.3	52.3
March	1.9	67.5	47.3	57.4
April	1.04	73.7	50.9	62.3
May	0.42	82.7	57.6	70.2
June	0.24	91.4	63.9	77.6
July	0.03	97.7	69.3	83.5
August	0	96.5	67.9	82.2
September	0.05	90.7	63.4	77.1
October	0.56	78.7	54.6	66.7
November	0.87	64.9	45.4	55.1
December	1.79	55.3	39.8	47.5
Annual	10.99	76.3	53.7	65

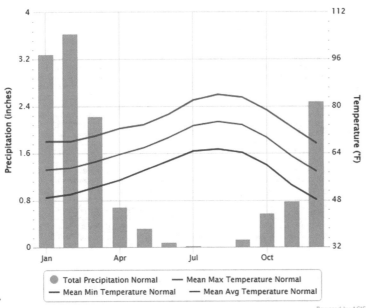

Monthly Climate Normals (1991–2020) – Los Angeles Downtown Area, CA (ThreadEx)

Month	Total precipitation normal (in.)	Mean max temperature normal (°F)	Mean min temperature normal (°F)	Mean avg temperature normal (°F)
January	3.29	68	48.9	58.4
February	3.64	68	50	59
March	2.23	69.9	52.4	61.1
April	0.69	72.4	54.8	63.6
May	0.32	73.7	58.1	65.9
June	0.09	77.2	61.4	69.3
July	0.02	82	64.7	73.3
August	0	84	65.4	74.7
September	0.13	83	64.2	73.6
October	0.58	78.6	59.9	69.3
November	0.78	72.9	53.1	63
December	2.48	67.4	48.2	57.8
Annual	14.25	74.8	56.8	65.8

Monthly Climate Normals (1991–2020) – PALM SPRINGS ASOS, CA

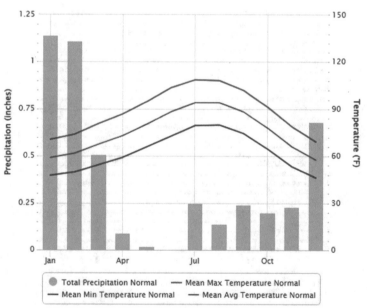

Powered by ACIS

Month	Total precipitation normal (in.)	Mean max temperature normal (°F)	Mean min temperature normal (°F)	Mean avg temperature normal (°F)
January	1.14	70.5	47.6	59
February	1.11	73.7	49.7	61.7
March	0.51	80.6	54.4	67.5
April	0.09	86.7	59.1	72.9
May	0.02	94.7	65.9	80.3
June	0	103.6	72.7	88.2
July	0.25	108.6	79.4	94
August	0.14	108.1	79.8	94
September	0.24	101.8	74.4	88.1
October	0.2	91.1	64.5	77.8
November	0.23	78.7	53.4	66
December	0.68	69.2	46.2	57.7
Annual	4.61	88.9	62.3	75.6

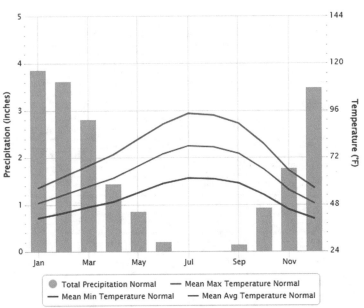

Monthly Climate Normals (1991–2020) – Sacramento Downtown Area, CA (ThreadEx)

Month	Total precipitation normal (in.)	Mean max temperature normal (°F)	Mean min temperature normal (°F)	Mean avg temperature normal (°F)
January	3.66	56	39.2	47.6
February	3.49	61.3	41.5	51.4
March	2.68	66.3	44.5	55.4
April	1.26	72.1	47	59.5
May	0.75	80.3	52	66.1
June	0.23	87.9	56.5	72.2
July	0	92.6	59.2	75.9
August	0.04	91.9	58.8	75.3
September	0.09	88.5	56.5	72.5
October	0.85	78.8	50.3	64.5
November	1.66	65	42.7	53.9
December	3.43	56	38.5	47.3
Annual	18.14	74.7	48.9	61.8

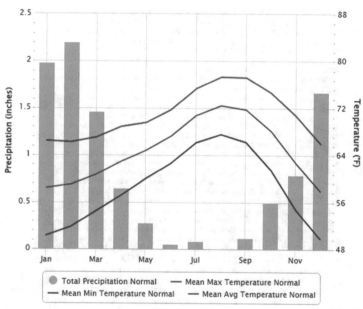

Monthly Climate Normals (1991–2020) – SAN DIEGO INTERNATIONAL AP, CA

Powered by ACIS

Month	Total precipitation normal (in.)	Mean max temperature normal (°F)	Mean min temperature normal (°F)	Mean avg temperature normal (°F)
January	1.98	66.4	50.3	58.4
February	2.2	66.2	51.8	59
March	1.46	67	54.5	60.7
April	0.65	68.8	57.1	62.9
May	0.28	69.5	60	64.8
June	0.05	71.7	62.6	67.2
July	0.08	75.3	66.1	70.7
August	0.01	77.3	67.5	72.4
September	0.12	77.2	66.2	71.7
October	0.5	74.6	61.5	68.1
November	0.79	70.7	54.8	62.7
December	1.67	66	49.8	57.9
Annual	9.79	70.9	58.5	64.7

Monthly Climate Normals (1991–2020) – SAN FRANCISCO INTERNATIONAL AP, CA

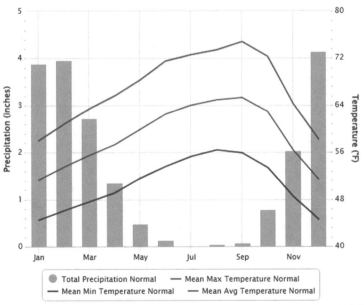

Powered by ACIS

Month	Total precipitation normal (in.)	Mean max temperature normal (°F)	Mean min temperature normal (°F)	Mean avg temperature normal (°F)
January	3.89	58	44.5	51.3
February	3.96	60.8	46.1	53.5
March	2.73	63.4	47.6	55.5
April	1.36	65.6	49.1	57.3
May	0.48	68.3	51.6	59.9
June	0.14	71.5	53.6	62.5
July	0	72.6	55.3	64
August	0.04	73.4	56.4	64.9
September	0.07	74.8	55.9	65.3
October	0.79	72.3	53.4	62.9
November	2.04	64.3	48.5	56.4
December	4.14	58.2	44.6	51.4
Annual	19.64	66.9	50.6	58.7

Appendix B: Historical Winter Storms

The following is a list, not complete, of several significant winter storms:

The Great Flood of 1861–62. William H. Brewer (1930) of the Whitney California Geological Survey wrote from San Francisco on Sunday, January 19, 1862, "The amount of rain that has fallen is unprecedented in the history of the state. The great central valley of the state is under water – the Sacramento and San Joaquin valleys - a region of 250 to 300 miles long and an average of twenty miles wide, a district of five thousand or six thousand square miles, or probable an area of three to three and a half millions of acres!" (Source: Goodridge 1996).

Storm of August 12, 1891 at Campo weather station southeast of San Diego recorded 16.10″ in 24-h. which was one of the most intense storms ever recorded in the United States and would have had a return period of one in ten million years.

March 1907 Continuous downpours flooded the Sacramento Valley turning most of the Valley into one large lake. This was followed by heavy wet snow, over 40″ in Susanville.

January 1909. Continuous heavy rainfall from Fort Ross on the coast to Greenville on the Feather River Basin. La Porte had 57.41 in. in 20 days which was 5.38 standard deviations above the mean representing a return period of 12,000 years. The 1907 and 1909 floods finally changed flood control strategies from levies to more formidable structures.

The Tehachapi Storm of September 28 to October 1, 1932. The Tehachapi Storm resulted from a tropical storm that moved up the Gulf of California and broke up over the Tehachapi Mountains. It caused a million dollars in damages and cost 15 people their lives. Tehachapi had 7.11″ of rain, washing away 2 freight engines, tracks, and a service station.

1938 Los Angeles Flood. Over 10″ of rain was recorded at 16 different stations in southern California. Streamflows were some of the highest ever recorded in southern California. 87 lives were lost and $78 million in damages occurred in this flood. This was also the last straw for the city following several damaging floods and led to the building of the concrete LA River.

The Christmas Storm of December 1955. A warm storm from December 17 to 27 melted snow in the Central Sierras, while 19 stations in the upper Sacramento River Valley reported daily rainfall over 10″ as rain continued all week leading to extensive flooding.

Winter of 1969. Some of the most severe flooding since 1938 took place in southern California. Over 200 stations in the region had the highest 60 consecutive day's totals ever from January 13 to March 13, 1969. The highest daily rainfall, 24.92″, the second highest in the state, took place at Lytle Creek Power House in the San Gabriel Mountains.

El Niño 1982–83 storms. According to Goodrige (1992), it had been 93 years since the state had as much rain as in 1983. Forty-five stations had annual rainfall over the calculated one in a thousand year amounts for 1983. These records extended from the Klamath River Basin in the north to Borrego Desert to the far south.

March 9–10, 1995, winter storm. Heavy rains led to flooding on the Napa and Russian Rivers on the 9th, with more flooding on the 10th on the Salinas River and the San Joaquin Valley. Over the two days, there was a total of 16 deaths and 1.1 billion dollars in property damage.

1997 New Year's Flood. Rains and snow since late December 1996 led to widespread floods and landslides throughout northern California. The Yosemite Valley flooded for the first time in over a century.

2016–2017 winter storms. After six years of drought, the storms came and didn't stop bringing record snows to the Sierras and flooding rains, with the worst flooding in northern California. Atmospheric rivers were to blame.

Information on these and other winter storms are from the following sources:

California Department of Water Resources. (2022, December 5). *Flood*. Dept. of Water Resources. https://water.ca.gov/Water-Basics/Flood

Dettinger, M. D., Ralph, F. M., Das, T., Neiman, P. J., & Cayan, D. R. (2011). Atmospheric rivers, floods and the water resources of California. *Water, 3*, 445–478. https://www.mdpi.com/2073-4441/3/2/445

Goodridge, J. D. (1992). A study of 1000 year storms in California. In Ninth Annual PACLIM Workshop, Pacific Grove, CA.

Goodridge, J. D. (1996). Data on California's extreme rainfall 1862–1995. In *1996 California Weather Symposium*, Rocklin, CA.

Historical California Winter Storms. (2018, February 27). NBC News. https://www.nbclosangeles.com/news/local/Historic-California-Storms-Flooding-Rainfall-Damage-Winter-411456905.html

Lloyd, J. (2021, October 25). *Historical California winter storms: Devastating floods, landslides, dust storms*. https://www.nbclosangeles.com/weather-news/historic-california-storms-flooding-rainfall-damage-winter/19208/

The nightmare California flood more dangerous than a huge earthquake. *LA Times*, March 25, 2018. https://www.latimes.com/local/california/la-me-california-flood-20180325-htmlstory.html

Appendix C: Notable California Fires

Information below are from the Cal Fire websites:

https://www.fire.ca.gov/media//4jandlhh/top20_acres.pdf;

https://www.fire.ca.gov/media/lbfd0m2f/top20_deadliest.pdf

Top 20 California wildfires

Fire name (Cause)	Date	County	Acres	Structures	Death
1 AUGUST COMPLEX (Lightning)	Aug. 2020	Mendocino, Humboldt, Trinity, Tehama, Glenn, Lake, and Colusa	1,032,648	935	1
2 DIXIE (Powerlines)	Jul. 2021	Butte, Plumas, Lassen, Shasta and Tehama	963,309	1311	1
3 MENDOCINO COMPLEX (Human related)	Jul. 2018	Colusa, Lake, Mendocino and Glenn	459,123	280	1

(continued)

© The Editor(s) (if applicable) and The Author(s), under exclusive license to Springer Nature Switzerland AG 2023
S. LaDochy and M. Witiw, *Fire and Rain*,
https://doi.org/10.1007/978-3-031-32273-0

(continued)

Top 20 California wildfires					
Fire name (Cause)	Date	County	Acres	Structures	Death
4 SCU LIGHTNING COMPLEX (Lightning)	Aug. 2020	Stanislaus, Santa Clara, Alameda, Contra Costa, and San Joaquin	396,625	225	0
5 CREEK (Undetermined)	Sept. 2020	Fresno and Madera	379,895	858	0
6 LNU LIGHTNING COMPLEX (Lightning/arson)	Aug. 2020	Napa, Solano, Sonoma, Yolo, Lake and Colusa	363,220	1491	6
7 NORTH COMPLEX (Lightning)	Aug. 2020	Butte, Plumas and Yuba	318,935	2352	15
8 THOMAS (Powerlines)	Dec. 2017	Ventura and Santa Barbara	281,893	1060	2
9 CEDAR (Human related)	Oct. 2003	San Diego	273,246	2820	15
10 RUSH (lightning)	Aug. 2012	Lassen	273,246	0	0
11 RIM (Human related)	Aug. 2013	Tuolumne	257,314	112	0
12 ZACA (Human related)	Jul. 2007	Santa Barbara	240,207	1	0
13 CARR (Human related)	Jul. 2018	Shasta County and Trinity	229,651	1614	8
14 MONUMENT (Lightning)	Jul. 2021	Trinity	223,124	28	0
15 CALDOR (Human related)	Aug. 2021	Alpine, Amador, and El Dorado	221,835	1005	1
16 MALTILIJA (Undetermined)	Sept. 1932	Ventura and Santa Barbara	220,000	0	0
17 RIVER COMPLEX (Lightning)	Jul. 2021	Siskiyou and Trinity	199,359	122	0
18 WITCH (Powerlines)	Oct. 2007	San Diego	197,990	1650	2
19 KLAMATH THEATRE COMPLEX (Lightning)	Jun. 2008	Siskiyou	192,038	0	2

(continued)

(continued)

Top 20 California wildfires					
Fire name (Cause)	Date	County	Acres	Structures	Death
20 MARBLE CONE (Lightning)	Jul. 1977	Monterey	177,866	0	0

There is no doubt that there were fires with significant acreage burned in years prior to 1932, but those records are less reliable, and this list is meant to give an overview of the large fires in more recent times

This list does not include fire jurisdiction. These are the Top 20 regardless of whether they were state, federal, or local responsibility

Numbers not final

Source Cal Fire. https://www.fire.ca.gov/our-impact/statistics

Top 20 deadliest California wildfires					
Name	Date	Counties	Acres	Structures	Deaths
1 CAMP	Nov. 2018	Butte	153,336	18,804	85
2 GRIFFITH PARK	Oct. 1933	Los Angeles	47	0	29
3 TUNNEL—Oakland Hills	Oct. 1991	Alameda	1600	2900	25
4 TUBBS	Oct. 2017	Napa, Sonoma	36,807	5636	22
5 NORTH COMPLEX	Aug. 2020	Butte, Plumas, Yuba	318,937	2352	15
6 CEDAR	Oct. 2003	San Diego	273,246	2820	15
7 RATTLESNAKE	Jul. 1953	Glenn	1340	0	15
8 LOOP	Nov. 1966	Los Angeles	2028	0	12
9 HAUSER CREEK	Oct. 1943	San Diego	13,145	0	11
10 INAJA	Nov. 1956	San Diego	43,904	0	11
11 IRON ALPS COMPLEX	Aug. 2008	Trinity	105,855	10	10
12 REDWOOD VALLEY	Oct. 2017	Mendocino	36,523	543	9
13 HARRIS	Oct. 2007	San Diego	90,440	548	8

(continued)

(continued)

Top 20 deadliest California wildfires

Name	Date	Counties	Acres	Structures	Deaths
14 CANYON	Aug. 1968	Los Angeles	22,197	0	8
15 CARR	Jul. 2018	Shasta, Trinity	229,651	1614	7
16 LNU Lightning Complex	Aug. 2020	Napa/ Sonoma/ Yolo/ Stanislaus, Lake	363,220	1491	6
17 ATLAS	Oct. 2017	Napa, Solano	51,624	781	6
18 OLD	Oct. 2003	San Bernardino	91,281	1003	6
19 DECKER	Aug. 1959	Riverside	1425	1	6
20 HACIENDA	Sept. 1955	Los Angeles	1150	0	6

Numbers not final

Fires with the same death count are listed by most recent. Several fires had 6 fatalities, but only the most recent are listed

This list does not include fire jurisdiction. These are the Top 20 regardless of whether they were state, federal, or local are listed

Source Cal Fire. https://www.fire.ca.gov/our-impact/statistics

Tunnel Fire

Northern CA has the Diablo wind, a hot, dry NE wind that sweeps down the hills east of SF through the canyons of the Diablo Range, which runs N–S on the east side of the SF Bay. While the devastating Oakland Tunnel Fire of 1991 was mentioned in Chap. 1, one of the worst firestorms in CA history, caused mainly by a Diablo wind, burned the Oakland and Berkeley Hills east of SF on Sept. 17, 1923. Within 3 h, the fire scorched 577 homes. It was Berkeley's worst natural disaster.

Witch Fire

In October 2007, a series of wildfires burned some 500,000 acres of land and destroyed 1500 homes across southern California. The fires stretched from Santa Barbara County to the US–Mexico border, and seven counties were declared disaster areas. As firefighters fought to control the blazes, the fires were helped by unusually hot weather and extreme drought in southern

California. However, one of the biggest and most challenging factors in the fight against the fires was the wind. In the fall of 2007, the annual Santa Ana winds blew stronger than usual, fanning the flames to greater ferocity.

Causes: Persistent drought, Santa Ana winds, volative native, and exotic vegetation.

Damages: ~ 300,000 ha. burnt, 22 deaths, 3570 homes destroyed.

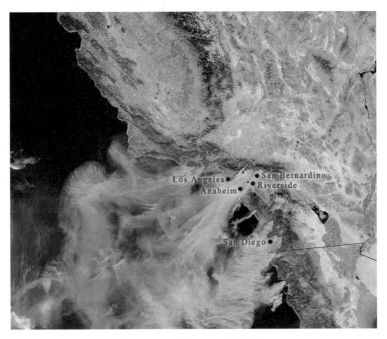

2003 Witch Fire (NASA)

Other Santa Ana wind-driven fires include:

During a 10-day period in September 1970, Santa Ana winds fanned fires that destroyed nearly a half million acres. Property damage during fires that burned 80,000 acres in late November 1980 was estimated at $40 million.

In 1993, 26 major fires fanned by Santa Ana winds killed four people, destroyed or damaged over 1200 structures, and resulted in almost $1 billion in damage.

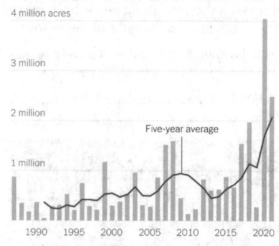

Acres burned in California wildfires since 1987

Data for 2020 and 2021 are estimates. Source: California Department of Forestry and Fire Protection

Appendix D: The American Lung Association 2022 "State of the Air"

Here are the rankings for each category:

Ozone

1. Los Angeles-Long Beach, California
2. Bakersfield, California
3. Visalia, California
4. Fresno-Madera-Hanford, California
5. Phoenix-Mesa, Arizona
6. San Diego-Chula Vista-Carlsbad, California
7. Denver-Aurora, Colorado
8. Houston-The Woodlands, Texas
9. Sacramento-Roseville, California
10. Salt Lake City-Provo-Orem, Utah.

Year-Round Particle Pollution

1. Bakersfield, California
2. Fresno-Madera-Hanford, California
3. Visalia, California
4. San Jose-San Francisco-Oakland, California
5. Los Angeles-Long Beach, California
6. Medford-Grants Pass, Oregon

7. Fairbanks, Alaska
8. Phoenix-Mesa, Arizona
9. Chico, California
10. El Centro, California.

Short-Term Particle Pollution

1. Fresno-Madera-Hanford, California
2. Bakersfield, California
3. Fairbanks, Alaska
4. San Jose-San Francisco-Oakland, California
5. Redding-Red Bluff, California
6. Chico, California
7. Sacramento-Roseville, California
8. Los Angeles-Long Beach, California
9. Yakima, Washington
10. Visalia, California.

For latest listings: https://www.lung.org/research/sota/city-rankings/most-polluted-cities.

Appendix E: Two Years Before the Mast, by Richard Henry Dana Jr., 1840

The following are excerpts from the book describing southern California weather events between 1834 and 1836.

This wind (the south-easter) is the bane of the coast of California. Between the months of November and April, (including a part of each,) which is the rainy season in this latitude, you are never safe from it, and accordingly, in the ports which are open to it, vessels are obliged, during these months, to lie at anchor at a distance of three miles from the shore, with slip-ropes on their cables, ready to slip and go to sea at a moment's warning. The only ports which are safe from this wind are San Francisco and Monterey in the north, and San Diego in the south.

Santa Barbara-In the first place, it was a beautiful day, and so warm that we had on straw hats, duck trowsers, and all the summer gear; and as this was mid-winter, it spoke well for the climate; and we afterwards found that the thermometer never fell to the freezing-point throughout the winter, and that there was very little difference between the seasons, except that during a long period of rainy and south-easterly weather, thick clothes were not uncomfortable…hills have no large trees upon them, they having been all burnt by a great fire which swept them off about a dozen years before, and they had not yet grown up again. The fire was described to me by an inhabitant, as having been a very terrible and magnificent sight. The air of the whole valley was so heated that the people were obliged to leave the town and take up their quarters for several days upon the beach.

Dana, R. H., Jr. (1840). *Two years before the mast*. Available at Project Gutenberg. https://gutenberg.org/files/2055/2055-h/2055-h.htm

Index